PELICAN BOOKS

COMPUTER POWER AND HUMAN REASON

Joseph Weizenbaum is Professor of Computer Science and a member of the Laboratory of Computer Science at the Massachusetts Institute of Technology in the U.S.A. At the beginning of his career with computers, circa 1950, he worked on the Bush Differential Analyzer, an analogue computer, and helped to design and build a digital computer at Wayne University in Detroit, Michigan. In 1955, after a journeyman career as a programmer-analyst, he became a member of the General Electric team which designed and built the first computer system dedicated to banking operations. Among his technical contributions are the list processing system SLIP and the natural language understanding program ELIZA. Professor Weizenbaum has held academic appointments at Harvard University, the Technical University of Berlin and the University of Hamburg in Germany. In 1973 he was a Fellow of the Center for Advanced Studies in the Behavioral Sciences, at Stanford, California. He is a Fellow of the American Association for the Advancement of Science, a member of the New York Academy of Science and of the European Academy of Science. He is also a member of the National Advisory Council of the Fellowship of Reconciliation, the American branch of an international peace group, and of the Advisory Committee for Disarmament Programs of the American Friends Service Committee.

JOSEPH WEIZENBAUM

COMPUTER POWER AND HUMAN REASON

FROM JUDGMENT TO CALCULATION

PENGUIN BOOKS

Penguin Books Ltd, Harmondsworth, Middlesex, England
Penguin Books, 40 West 23rd Street, New York, New York 10010, U.S.A.
Penguin Books Australia Ltd, Ringwood, Victoria, Australia
Penguin Books Canada Ltd, 2801 John Street, Markham, Ontario, Canada L3R 1B4
Penguin Books (N.Z.) Ltd, 182–190 Wairau Road, Auckland 10, New Zealand

First published by W. H. Freeman and Company 1976
Published with a new preface in Pelican Books 1984

Made and printed in Great Britain by
Richard Clay (The Chaucer Press) Ltd,
Bungay, Suffolk

to Ruth

CONTENTS

PREFACE

to the First Edition

This book is only nominally about computers. In an important sense, the computer is used here merely as a vehicle for moving certain ideas that are much more important than computers. The reader who looks at a few of this book's pages and turns away in fright because he spots an equation or a bit of computer jargon here and there should reconsider. He may think that he does not know anything about computers, indeed, that computers are too complicated for ordinary people to understand. But a major point of this book is precisely that we, all of us, have made the world too much into a computer, and that this remaking of the world in the image of the computer started long before there were any electronic computers. Now that we have computers, it becomes somewhat easier to see this imaginative transformation we have worked on the world. Now we can use the computer itself—that is the idea of the computer—as a metaphor to help us understand what we have done and are doing.

We are all used to hearing that the computer is a powerful new instrument. But few people have any idea where the power of a computer comes from. Chapters I to III are devoted to explaining just that. With a modest investment in time and intellectual energy, anyone who can read this Preface should be able to work his way through those chapters. Chapters II and III will be the most difficult, but the reader who cannot master them should not therefore abandon the rest of the book. Really, the only point Chapters II and III make is that computers are in some sense "universal" machines, that they can (in a certain sense which is there explained) do "anything." The reader who is willing to take that assertion on faith may well wish to skip from Chapter I (which he should read) to Chapter IV. Perhaps after he has finished the whole book, he will be tempted to try Chapters II and III again.

The rest of the book contains the major arguments, which are in essence, first, that there is a difference between man and machine, and, second, that there are certain tasks which computers *ought* not be made to do, independent of whether computers *can* be made to do them.

The writing of this book has been an adventure to me. First and most important, I have been cheered beyond my power to say by the generosity and the intellectual and emotional support given me by people who owe me absolutely nothing. But now I am very greatly in their debt. I am thinking primarily of Lewis Mumford, that grand old man, of Noam Chomsky, and of Steven Marcus, the literary critic. Each of them read large sections of the manuscript in preparation (Lewis Mumford read all of it) and contributed the wisest and most useful kinds of criticism. But more than that, each encouraged me to go on when I despaired. For there was often cause for despair. I am acutely aware, for example, that there is nothing I say in this book that has not been said better, certainly more eloquently, by others. But, as my friends continued to point out to me, it seemed important to say these things again and again. And, as Lewis Mumford often remarked, it sometimes matters that a member of the scientific establishment say some things that humanists have been shouting for ages.

More formally, I am indebted to my University, the Massachusetts Institute of Technology, for granting me a leave of two years so that I might pursue first the thinking which preceded the writing, then the writing itself. I spent the first of those two happy years at the Center for Advanced Study in the Behavioral Sciences at Stanford, California.* It was there that I met Steven Marcus, as well as others of the Center's Fellows, who struggled mightily to educate a primitive engineer. The names John Platt, Paul Armer, Herbert Weiner, Fredrick Redlich, Alexander Mitscherlich, and Israel Scheffler immediately come to mind. I thank them for their efforts. No failure of mine should, however, be counted against any of them. I spent the second year as a Vinton Hayes Research Fellow at Harvard University. There I had the good fortune to be able to renew an old collegial association, namely, one with Professor Thomas Cheatham, an outstanding computer scientist. He took the trouble to read almost the whole manuscript as it sprang from my pen. Professor Hilary Putnam of Harvard's Philosophy Department gave me many hours of his valuable time. Without his help, encouragement, and guidance I would have fallen into many more traps than I actually did. It was also a stroke of good luck that Daniel C. Dennett, an outstanding young philosopher from Tufts University, happened to be spending the year at Harvard just when I was there. His patience with my philosophical naivety was unlimited. I can never adequately discharge my debts to all these good people.

These few words of thanks acknowledge the fact that this book—like, I suspect, most others—has many co-authors whose names will not appear on its cover. But in this instance that confession would be grievously incomplete if it did not include an acknowledgment of the critical contributions that the book's manuscript editor, Aidan Kelly, made to it. I cannot, in a few words, summarize what he did. Perhaps readers will understand if I say simply that Aidan Kelly is a poet.

Finally, everyone who has ever written a book will know what an enormous burden such a task imposes on the author's family. My wife, Ruth, suffered my retreats to my study with the utmost

* My fellowship was supported in part by National Science Foundation Grant No. SSH71-01834 A01 from the Research Applied to National Needs (RANN) Program of NSF to the Center. Of course, the opinions, findings, conclusions, and recommendations contained herein are entirely mine and do not necessarily reflect the views of any sponsor.

good will and patience. She helped me over the inevitable bouts
with the feelings of guilt that overcome an author when he is writing—for then he is not with his family even when he is with them—
and when he is not writing—for then he is not doing what he has set
himself to do. My children counted the pages as they mounted on
my desk. And they grieved when, as often happened, the stack of
pages in the wastebasket grew more quickly than that on my desk.
Most of all, they cheerfully endured the endless progress reports
that punctuated our dinner-table conversation. This book is Ruth's
and our children's as much as it is mine.

Fall 1975   *Joseph Weizenbaum*
Cambridge, Massachusetts

PREFACE
to the 1984 Penguin Edition

The publication of *Computer Power and Human Reason* in a Penguin edition gives me opportunity and occasion to think about the time that has passed since the original American edition was published by Freeman and Company in 1976.

One question to ask is whether the book has made any difference in the world. I've often thought that I now hear very much less of the euphoric claiming and predicting from the artificial intelligence (AI) community than before the book's publication. Perhaps my criticisms of the AI community had that effect. On the other hand, perhaps AI simply matured a little. I am more nearly certain that the book encouraged many people to believe in their own ideas, and that that remains its principal merit. It is really extraordinarily difficult for most people to hold on to an idea in the absence of credible evidence that is shared by others. But to know that even only a single other person has a thought I have confirms and validates that thought. In this respect there is a much more important difference between one and two persons than between two and, say, ten. I know this from the very many readers' letters I have received over the years, letters which express their writers' gratitude for having, in a sense, their reason confirmed. We appear to have come to a time in which the ideas that there are differences between human beings and machines, that there are experiences that human beings can have but machines cannot, and that therefore (at least!) there are thoughts humans but not machines can have, we have come to a time in which the holding of such ideas is a lonely preoccupation, a business that tempts one to doubt of one's own sanity. In the book I wrote that the necessity to debate such ideas at all is an index to the insanity of our time. The danger now is that the debate will stop—not because the absurdity of equating human beings with computers has finally been universally recognized, but because the voices defending human-centered positions are becoming ever fainter.

Much has happened in the world since *Computer Power and Human Reason* first appeared in American book stores in 1976. I think it is universally agreed that the world has become a much more dangerous place during that time. On the whole, the technological fixes which were advertised as instruments which would give us greater international stability and security have in fact contributed mainly to destabilization and greater insecurity. The press of military necessity, among other factors, yielded ever smaller electronic components and almost miraculous bulk manufacturing techniques—very large scale integration (VLSI), for example. Physically smaller but functionally more powerful computers in missiles allow their "payload," that is their death dealing portion, to be increased while at the same time "improving" their aiming accuracy. (The architect of the German V-2 rockets, Werner von Braun, entitled his autobiography *I Aim For the Stars*. Londoners, however, will remember that he usually hit London. Victims of modern war may have the satisfaction of knowing that the bombs which destroy their cities are *intended* to destroy them.) An incidental fallout of such "progress" is that computers in general became smaller, faster, more powerful, and cheaper. It is this side-effect which is trumpeted by the media and which dominates the attention of the general public.

I certainly failed to foresee the near unanimity and the enormous magnitude of the enthusiasm with which the public—at least in America—welcomed the computer into its day to day affairs. Citizens' wallets bulge with ever more magnetically striped plastic cards, each of which, upon being read by a machine, admits its bearer into some routine activity, perhaps a ride on a public transport vehicle or a withdrawal of money from a banking machine, which was, only a short time ago, mediated by people. Most Americans now own computers of one sort or another. Micro computers inhabit washing machines and cameras, automobile engines and watches, typewriters and telephones. Most of us are as unaware of these micros as we are of the many small electric motors which are also all around us. Electronic hand calculators, most of them very cheap, also abound. However, the full-fledged home computer, that is, a free-standing, full-scale, programmable machine with its own input/output devices, while available and widely used by business, has not yet become as ubiquitous a household possession as, say, the telephone.

The kind of computers that have slipped into the consciousness of the general public are the game-playing computers in the arcades and those which couple to television sets. Computer games certainly existed at the time I wrote the book. In retrospect, it is obvious that precisely those that were most popular at M.I.T. and at other universities, for example the game "Space War," were the forerunners of the games which now dominate in arcades and homes. Why I missed their significance eight years ago is not at all clear to me.

There is no question in my mind now that the computer-game phenomenon, at least as it is playing itself out in the United States, powerfully validates the general cultural pessimism which *Computer Power and Human Reason* expresses. Whatever despair our society's use of television induces, it must be doubled and redoubled by the vision of countless youngsters standing hypnotized before computer displays, their hands moving in the manner of those of a shell-shocked soldier. I think what is happening to young people in the computer arcades and in some of their class-rooms is a parable of our time, a sad and disturbing story.

To understand the content of most computer games, one has only to sample the main stream of (at least American) television to know that it consists mainly of what would in German be called *Unsinn, Bloedsinn, Wahnsinn*, that is, nonsense, stupidity, and insanity. Inane so called situation comedies relieve the otherwise almost constant stream of mindless violence. On Saturday and Sunday mornings, the same fare is presented yet again, only now in cartoon form so that children can absorb it even more directly. The mass of computer and arcade games present precisely the same fare translated, of course, to take advantage of the new medium. However, whereas the television viewer passively receives, the computer game player actively participates. In concrete terms, this means that, while the television viewer watches, say, U-boat commanders launch torpedos against "enemy" ships and shout with joy as their targets disappear beneath the waves, the computer game player launches torpedos and himself *experiences* the thrill of the torpedo run. I almost wrote "the thrill of *killing*," and an important point hangs on this near slip of the pen: I dare say very few actual submarine captains experience *killing* in connection with pushing a button that initiates a torpedo's rush towards its target, nor do bombardiers in airplanes have that experience when they launch their bombs. Most human beings would be incapable of such actions if they were not able to maintain what a physician might call a "clinical distance" from the ultimate consequences of their actions. A less euphemistic way of saying the same thing is that much intensive training in psychic numbing is required before an ordinary person can launch torpedos that sink ships or release bombs that vaporize people several miles below.

The world's military and naval establishments do indeed devote much time and energy to training regimens which abstract from acts of killing all but the most innocuous, apparently innocent, technical operations. Similar remarks apply to people who produce components of weapon systems or ever more clever means of guiding missiles unerringly to the targets intended for them by their masters. There is an eager market in this world for people who are already psychically numb by the time they enter the world's workforces, that is, who are superbly trained to make no connections between what they do and the ultimate effect of what they do on what might be called the end users of the product of their labors, in other words, on their final victims.

Most of the computer and arcade games I have seen are trainers for just that skill. Space ships and airplanes are shot down in great numbers. Megaton bombs shower whole countries. Many games are so constructed that there can be no survivors: "winning" consists of keeping the game going longer than has any other player, and that generally means being a quicker and more effective killer. There is even a computer game on the American market called "Custer's Revenge," in which that player wins who has *raped* the most American Indian women! In that game, the most gruesome and frightful insult a man can wreak upon a woman is cleansed of all torment, horror, anguish. Only abstract operations on plastic buttons remain. This is what I mean by psychic numbing.

At this writing, the American educational establishment is in an almost panic-like state. The mass media have declared us to be the "Information Society." The pressure on schools to provide computer instruction to children of age five years or older is all but irresistible. America seems to be in the grip of a fad which recalls the CB radio, Rubik's Cube, and the Hula Hoop madnesses of years past. The effects of the current disorder, however, are not likely to be as trivial as were those of the former. The minds of our youth are to be shaped in new ways. This is not mere entertainment. Is this at all related to the computer-game phenomenon I have just been discussing? Yes, in at least two ways.

First, in that the vast majority of the small personal computers, that is, those intended for the direct personal use of students, already installed in schools, come prepared to play many of the most popular arcade games. For many youngsters, these games are what lure them to the computer in the first place. In other words, we may expect the "education" in violence without guilt and in psychic numbing, the training offered by the arcade and often by the home computer as well, to be continued and reinforced in the school and by whatever little moral authority the schools may have left.

Second, in that much of the normal, ordinary computing that children will presumably be taught by way of preparing them for their roles in the Information Society, such things as model building, for example, has the same potential of separating what one does from the effects of what one does as do the computer games I have just discussed, only the process is more subtle. It is relatively easy to avoid contaminating young minds, even while instructing them in the design and use of computational models, to be sure. All that we require are alert teachers who themselves understand what models are and what one can and cannot learn from them.

I discussed models and theories (Chapter V) in the book. What is important in the present context is that models embody only the *essential* features of whatever it is they are intended to represent. If a model of an automobile is intended for wind tunnel tests, then the outside shape of the model car is important, but no seats nor any other interior furnishings of the real automobile need be present in the model. What aspects of reality are and what are not embodied in a model is entirely a function of the model builder's purpose. But no matter what the purpose, a model, and here I am concerned especially with computer models of aspects of reality, must necessarily *leave out almost everything* that is actually present in the real thing. Hence what can be learned from manipulating the model is strictly limited *in every case*. Whoever knows and appreciates this fact, and keeps it in mind while teaching about the use of computers, has a chance to immunize his or her students against believing or making excessive claims for much of their computer work.

In reality, however, few teachers (at any level!) know and appreciate limitations of this kind. Nor is there much, if any, emphasis on such issues in most university computer-science curricula. Matters like these are deemed to be "merely" philosophical. Besides, the rush to flood the schools with computers precludes adequate training of teachers. The American educational establishment's response to the obvious impossibility of training teachers in numbers consistent with the extent of the flood of computers which it has loosed on the schools is to convert the disaster to a triumph: it is consequently *boasted* that in general American teachers of computer subjects know less of those subjects than do the youngsters they are charged with teaching.

An inevitable result of this condition is, I fear, that almost all youngsters who get their formal introduction to computers in primary and secondary schools will have been given a certain facility for the creation of computational models without any corresponding knowledge of their limitations. We will thus be raising new crops of young people fully prepared to foist ever more technological fixes on a world already overfull, glutted, intoxicated with them. But what is even more worrisome is that the kind of naive simplemindedness I am here talking about, results in the end in, yes *is*, an abdication of responsibility, a closing of the mind to reality without an accompanying sense of incompleteness; in other words, a psychic numbing to the ultimate consequences of one's work of the same kind as that induced in the computer arcade.

Earlier I invoked television as a model for the content of computer games. It will do for still further comparison. It seems to me that one cannot accurately determine, or even estimate, what television does to the viewer, if one restricts one's analysis to single viewers looking at their sets. The effect of television on a society is in large part determined by the phenomenon of millions of minds being constantly and simultaneously bathed by what television broadcasts. So it is also with computers. The effect of playing, say, the video game "Defender," in which cities are subjected to atomic bombardment, would be different were that game played by only one person, and then in isolation, than it is when it is played in a context already soaked through and through by the spirit, the ideology, of that and many similar games, all of which have been deeply internalized by the players' whole peer group. The effect on children of being exposed to the idolization of the kind of rationality required (at least for the present) to program computers, of the further elevation of calculation above judgment, must be understood in the context of the culture in which it takes place. Our culture, as I tried to say in the original book, is already nearly committed to the proposition that the only legitimate knowledge we can gain of our world is that yielded by science. All thinking, dreaming, feeling, indeed all other sources of insight have already been delegitimated. The indoctrination of our children's minds with simplistic and uninformed computer idolatry, and that is almost certainly what most of computer instruction is and will be, is a pandemic phenomenon. This alone seems to me to validate the cultural pessimism expressed in the original edition of this book.

Of course, I wish I could cheer the British reader with the message that events since the publication of the book in 1976 have falsified the basis of the despair with which some of its pages speak. Perhaps conditions are more civilized in the United Kingdom than in the U.S.A. If so, let this book be a warning.

Cambridge, Massachusetts, 1983   *Joseph Weizenbaum*

INTRODUCTION

In 1935, Michael Polanyi, then holder of the Chair of Physical Chemistry at the Victoria University of Manchester, England, was suddenly shocked into a confrontation with philosophical questions that have ever since dominated his life. The shock was administered by Nicolai Bukharin, one of the leading theoreticians of the Russian Communist party, who told Polanyi that "under socialism the conception of science pursued for its own sake would disappear, for the interests of scientists would spontaneously turn to the problems of the current Five Year Plan." Polanyi sensed then that "the scientific outlook appeared to have produced a mechanical conception of man and history in which there was no place for science itself." And further that "this conception denied altogether any intrinsic power to thought and thus denied any grounds for claiming freedom of thought."[1]

I don't know how much time Polanyi thought he would devote to developing an argument for a contrary concept of man and
history. His very shock testifies to the fact that he was in profound
disagreement with Bukharin, therefore that he already conceived of
man differently, even if he could not then give explicit form to his
concept. It may be that he determined to write a counterargument to
Bukharin's position, drawing only on his own experience as a scientist, and to have done with it in short order. As it turned out, however, the confrontation with philosophy triggered by Bukharin's rev-
elation was to demand Polanyi's entire attention from then to the
present day.

I recite this bit of history for two reasons. The first is to
illustrate that ideas which seem at first glance to be obvious and
simple, and which ought therefore to be universally credible once
they have been articulated, are sometimes buoys marking out
stormy channels in deep intellectual seas. That science is creative,
that the creative act in science is equivalent to the creative act in art,
that creation springs only from autonomous individuals, is such a
simple and, one might think, obvious idea. Yet Polanyi has, as have
many others, spent nearly a lifetime exploring the ground in which
it is anchored and the turbulent sea of implications which surrounds
it.

The second reason I recite this history is that I feel myself to
be reliving part of it. My own shock was administered not by any
important political figure espousing his philosophy of science, but
by some people who insisted on misinterpreting a piece of work I
had done. I write this without bitterness and certainly not in a defensive mood. Indeed, the interpretations I have in mind tended, if
anything, to overrate what little I had accomplished and certainly its
importance. No, I recall that piece of work now only because it
seems to me to provide the most parsimonious way of identifying
the issues I mean to discuss.

The work was done in the period 1964–1966, and was reported in the computer-science literature in January 1966 and August 1967. To summarize it briefly, I composed a computer program
with which one could "converse" in English. The human conversa-
tionalist partner would type his portion of the conversation on a

typewriter connected to a computer, and the computer, under control of my program, would analyze the message that had so been transmitted to it, compose a response to it in English, and cause the response to be typed on the computer's typewriter.

I chose the name ELIZA for the language analysis program because, like the Eliza of Pygmalion fame, it could be taught to "speak" increasingly well. Because conversations must be about something, that is, because they must take place within some context, the program was constructed in a two-tier arrangement, the first tier consisting of the language analyzer and the second of a script. The script is a set of rules rather like those that might be given to an actor who is to use them to improvise around a certain theme. Thus ELIZA could be given a script to enable it to maintain a conversation about cooking eggs or about managing a bank checking account, and so on. Each specific script thus enabled ELIZA to play a specific conversational role.

For my first experiment, I gave ELIZA a script designed to permit it to play (I should really say parody) the role of a Rogerian psychotherapist engaged in an initial interview with a patient. The Rogerian psychotherapist is relatively easy to imitate because much of his technique consists of drawing his patient out by reflecting the patient's statements back to him. The following conversation between a young lady and Eliza playing doctor illustrates both the Rogerian technique of encouraging a patient to keep talking and the operation of the computer program ELIZA. The first to "speak" is the young lady. The computer's responses are printed entirely in capitals.[2]

    Men are all alike.

    IN WHAT WAY

    They're always bugging us about something or other.

    CAN YOU THINK OF A SPECIFIC EXAMPLE

    Well, my boyfriend made me come here.

    YOUR BOYFRIEND MADE YOU COME HERE

    He says I'm depressed much of the time.

    I AM SORRY TO HEAR YOU ARE DEPRESSED

It's true. I am unhappy.

DO YOU THINK COMING HERE WILL HELP YOU NOT TO BE UNHAPPY

I need some help, that much seems certain.

WHAT WOULD IT MEAN TO YOU IF YOU GOT SOME HELP

Perhaps I could learn to get along with my mother.

TELL ME MORE ABOUT YOUR FAMILY

My mother takes care of me.

WHO ELSE IN YOUR FAMILY TAKES CARE OF YOU

My father.

YOUR FATHER

You are like my father in some ways.

WHAT RESEMBLANCE DO YOU SEE

You are not very aggressive but I think you don't want me to notice that.

WHAT MAKE YOU THINK I AM NOT VERY AGGRESSIVE

You don't argue with me.

WHY DO YOU THINK I DON'T ARGUE WITH YOU

You are afraid of me.

DOES IT PLEASE YOU TO BELIEVE I AM AFRAID OF YOU

My father is afraid of everybody.

WHAT ELSE COMES TO MIND WHEN YOU THINK OF YOUR FATHER

Bullies.

DOES THAT HAVE ANYTHING TO DO WITH THE FACT THAT YOUR BOYFRIEND MADE YOU COME HERE

DOCTOR, as ELIZA playing psychiatrist came to be known, soon became famous around the Massachusetts Institute of Technology, where it first came into existence, mainly because it was an easy program to demonstrate. Most other programs could not vividly demonstrate the information-processing power of a computer to visitors who did not already have some specialized knowledge, say, of some branch of mathematics. DOCTOR, on the other hand, could be appreciated on some level by anyone. Its power as a demonstration vehicle was further enhanced by the fact that the visitor could actually participate in its operation. Soon copies of DOCTOR, constructed on the basis of my published description of it, began appearing at other institutions in the United States. The program became nationally known and even, in certain circles, a national plaything.

The shocks I experienced as DOCTOR became widely known and "played" were due principally to three distinct events.

1. A number of practicing psychiatrists seriously believed the DOCTOR computer program could grow into a nearly completely automatic form of psychotherapy. Colby *et al.* write, for example,

> "Further work must be done before the program will be ready for clinical use. If the method proves beneficial, then it would provide a therapeutic tool which can be made widely available to mental hospitals and psychiatric centers suffering a shortage of therapists. Because of the time-sharing capabilities of modern and future computers, several hundred patients an hour could be handled by a computer system designed for this purpose. The human therapist, involved in the design and operation of this system, would not be replaced, but would become a much more efficient man since his efforts would no longer be limited to the one-to-one patient-therapist ratio as now exists."[3]*

I had thought it essential, as a prerequisite to the very possibility that one person might help another learn to cope with his emotional problems, that the helper himself participate in the other's experience of those problems and, in large part by way of his own em-

---

* Nor is Dr. Colby alone in his enthusiasm for computer administered psychotherapy. Dr. Carl Sagan, the astrophysicist, recently commented on ELIZA in *Natural History*, vol. LXXXIV, no. 1 (Jan. 1975), p. 10: "No such computer program is adequate for psychiatric use today, but the same can be remarked about some human psychotherapists. In a period when more and more people in our society seem to be in need of psychiatric counseling, and when time sharing of computers is widespread, I can imagine the development of a network of computer psychotherapeutic terminals, something like arrays of large telephone booths, in which, for a few dollars a session, we would be able to talk with an attentive, tested, and largely nondirective psychotherapist."

pathic recognition of them, himself come to understand them. There are undoubtedly many techniques to facilitate the therapist's imaginative projection into the patient's inner life. But that it was possible for even one practicing psychiatrist to advocate that this crucial component of the therapeutic process be entirely supplanted by pure technique—*that* I had not imagined! What must a psychiatrist who makes such a suggestion think he is doing while treating a patient, that he can view the simplest mechanical parody of a single interviewing technique as having captured anything of the essence of a human encounter? Perhaps Colby *et al.* give us the required clue when they write;

> "A human therapist can be viewed as an information processor and decision maker with a set of decision rules which are closely linked to short-range and long-range goals, ... He is guided in these decisions by rough empiric rules telling him what is appropriate to say and not to say in certain contexts. To incorporate these processes, to the degree possessed by a human therapist, in the program would be a considerable undertaking, but we are attempting to move in this direction."[4]

What can the psychiatrist's image of his patient be when he sees himself, as therapist, not as an engaged human being acting as a healer, but as an information processor following rules, etc.?

Such questions were my awakening to what Polanyi had earlier called a "scientific outlook that appeared to have produced a mechanical conception of man."

2. I was startled to see how quickly and how very deeply people conversing with DOCTOR became emotionally involved with the computer and how unequivocally they anthropomorphized it. Once my secretary, who had watched me work on the program for many months and therefore surely knew it to be merely a computer program, started conversing with it. After only a few interchanges with it, she asked me to leave the room. Another time, I suggested I might rig the system so that I could examine all conversations anyone had had with it, say, overnight. I was promptly bombarded with accusations that what I proposed amounted to spying on people's most

intimate thoughts; clear evidence that people were conversing with the computer as if it were a person who could be appropriately and usefully addressed in intimate terms. I knew of course that people form all sorts of emotional bonds to machines, for example, to musical instruments, motorcycles, and cars. And I knew from long experience that the strong emotional ties many programmers have to their computers are often formed after only short exposures to their machines. What I had not realized is that extremely short exposures to a relatively simple computer program could induce powerful delusional thinking in quite normal people. This insight led me to attach new importance to questions of the relationship between the individual and the computer, and hence to resolve to think about them.

3. Another widespread, and to me surprising, reaction to the ELIZA program was the spread of a belief that it demonstrated a general solution to the problem of computer understanding of natural language. In my paper, I had tried to say that no general solution to that problem was possible, i.e., that language is understood only in contextual frameworks, that even these can be shared by people to only a limited extent, and that consequently even people are not embodiments of any such general solution. But these conclusions were often ignored. In any case, ELIZA was such a small and simple step. Its contribution was, if any at all, only to vividly underline what many others had long ago discovered, namely, the importance of context to language understanding. The subsequent, much more elegant, and surely more important work of Winograd[5] in computer comprehension of English is currently being misinterpreted just as ELIZA was. This reaction to ELIZA showed me more vividly than anything I had seen hitherto the enormously exaggerated attributions an even well-educated audience is capable of making, even strives to make, to a technology it does not understand. Surely, I thought, decisions made by the general public about emergent technologies depend much more on what that public attributes to such technologies than on what they actually are or can and cannot do. If, as appeared to be the case, the public's attributions are wildly misconceived, then public decisions are bound to be misguided and

often wrong. Difficult questions arise out of these observations; what, for example, are the scientist's responsibilities with respect to making his work public? And to whom (or what) is the scientist responsible?

As perceptions of these kinds began to reverberate in me, I thought, as perhaps Polanyi did after his encounter with Bukharin, that the questions and misgivings that had so forcefully presented themselves to me could be disposed of quickly, perhaps in a short, serious article. I did in fact write a paper touching on many points mentioned here.[6] But gradually I began to see that certain quite fundamental questions had infected me more chronically than I had first perceived. I shall probably never be rid of them.

There are as many ways to state these basic questions as there are starting points for coping with them. At bottom they are about nothing less than man's place in the universe. But I am professionally trained only in computer science, which is to say (in all seriousness) that I am extremely poorly educated; I can mount neither the competence, nor the courage, not even the chutzpah, to write on the grand scale actually demanded. I therefore grapple with questions that couple more directly to the concerns I have expressed, and hope that their larger implications will emerge spontaneously.

I shall thus have to concern myself with the following kinds of questions:

1. What is it about the computer that has brought the view of man as a machine to a new level of plausibility? Clearly there have been other machines that imitated man in various ways, e.g., steam shovels. But not until the invention of the digital computer have there been machines that could perform intellectual functions of even modest scope; i.e., machines that could in any sense be said to be intelligent. Now "artificial intelligence" (AI) is a subdiscipline of computer science. This new field will have to be discussed. Ultimately a line dividing human and machine intelligence must be drawn. If there is no such line, then advocates of computerized psychotherapy may be merely heralds of an age in which man has finally been recognized as nothing but a clock-work. Then the con-

sequences of such a reality would need urgently to be divined and contemplated.

2. The fact that individuals bind themselves with strong emotional ties to machines ought not in itself to be surprising. The instruments man uses become, after all, extensions of his body. Most importantly, man must, in order to operate his instruments skillfully, internalize aspects of them in the form of kinesthetic and perceptual habits. In that sense at least, his instruments become literally part of him and modify him, and thus alter the basis of his affective relationship to himself. One would expect man to cathect more intensely to instruments that couple directly to his own intellectual, cognitive, and emotive functions than to machines that merely extend the power of his muscles. Western man's entire milieu is now pervaded by complex technological extensions of his every functional capacity. Being the enormously adaptive animal he is, man has been able to accept as authentically natural (that is, as given by nature) such technological bases for his relationship to himself, for his identity. Perhaps this helps to explain why he does not question the appropriateness of investing his most private feelings in a computer. But then, such an explanation would also suggest that the computing machine represents merely an extreme extrapolation of a much more general technological usurpation of man's capacity to act as an autonomous agent in giving meaning to his world. It is therefore important to inquire into the wider senses in which man has come to yield his own autonomy to a world viewed as machine.

3. It is perhaps paradoxical that just, when in the deepest sense man has ceased to believe in—let alone to trust—his own autonomy, he has begun to rely on autonomous machines, that is, on machines that operate for long periods of time entirely on the basis of their own internal realities. If his reliance on such machines is to be based on something other than unmitigated despair or blind faith, he must explain to himself what these machines do and even how they do what they do. This requires him to build some conception of their internal "realities." Yet most men don't understand computers to even the slightest degree. So, unless they are capable of very great skepticism (the kind we bring to bear while watching a stage magi-

cian), they can explain the computer's intellectual feats only by bringing to bear the single analogy available to them, that is, their model of their own capacity to think. No wonder, then, that they overshoot the mark; it is truly impossible to imagine a human who could imitate ELIZA, for example, but for whom ELIZA's language abilities were his limit. Again, the computing machine is merely an extreme example of a much more general phenomenon. Even the breadth of connotation intended in the ordinary usage of the word "machine," large as it is, is insufficient to suggest its true generality. For today when we speak of, for example, bureaucracy, or the university, or almost any social or political construct, the image we generate is all too often that of an autonomous machine-like process.

These, then, are the thoughts and questions which have refused to leave me since the deeper significances of the reactions to ELIZA I have described began to become clear to me. Yet I doubt that they could have impressed themselves on me as they did were it not that I was (and am still) deeply involved in a concentrate of technological society as a teacher in the temple of technology that is the Massachusetts Institute of Technology, an institution that proudly boasts of being "polarized around science and technology." There I live and work with colleagues, many of whom trust only modern science to deliver reliable knowledge of the world. I confer with them on research proposals to be made to government agencies, especially to the Department of "Defense." Sometimes I become more than a little frightened as I contemplate what we lead ourselves to propose, as well as the nature of the arguments we construct to support our proposals. Then, too, I am constantly confronted by students, some of whom have already rejected all ways but the scientific to come to know the world, and who seek only a deeper, more dogmatic indoctrination in that faith (although that word is no longer in their vocabulary). Other students suspect that not even the entire collection of machines and instruments at M.I.T. can significantly help give meaning to their lives. They sense the presence of a dilemma in an education polarized around science and technology, an education that implicitly claims to open a privileged access-path to fact, but that cannot tell them how to decide what is to count as fact. Even while they recognize the genuine importance of learning their craft, they rebel at working on projects that appear to address themselves neither to answering interesting questions of fact nor to solving problems in theory.

Such confrontations with my own day-to-day social reality have gradually convinced me that my experience with ELIZA was symptomatic of deeper problems. The time would come, I was sure, when I would no longer be able to participate in research proposal conferences, or honestly respond to my students' need for therapy (yes, that is the correct word), without first attempting to make sense of the picture my own experience with computers had so sharply drawn for me.

Of course, the introduction of computers into our already highly technological society has, as I will try to show, merely reinforced and amplified those antecedent pressures that have driven man to an ever more highly rationalistic view of his society and an ever more mechanistic image of himself. It is therefore important that I construct my discussion of the impact of the computer on man and his society so that it can be seen as a particular kind of encoding of a much larger impact, namely, that on man's role in the face of technologies and techniques he may not be able to understand and control. Conversations around that theme have been going on for a long time. And they have intensified in the last few years.

Certain individuals of quite differing minds, temperaments, interests, and training have—however much they differ among themselves and even disagree on many vital questions—over the years expressed grave concern about the conditions created by the unfettered march of science and technology; among them are Mumford, Arendt, Ellul, Roszak, Comfort, and Boulding. The computer began to be mentioned in such discussions only recently. Now there are signs that a full-scale debate about the computer is developing. The contestants on one side are those who, briefly stated, believe computers can, should, and will do everything, and on the other side those who, like myself, believe there are limits to what computers ought to be put to do.

It may appear at first glance that this is an in-house debate of little consequence except to a small group of computer technicians. But at bottom, no matter how it may be disguised by technological jargon, the question is whether or not every aspect of human thought is reducible to a logical formalism, or, to put it into the modern idiom, whether or not human thought is entirely computable. That question has, in one form or another, engaged thinkers in all ages. Man has always striven for principles that could organize and give sense and meaning to his existence. But before modern science fathered the technologies that reified and concretized its otherwise abstract systems, the systems of thought that defined man's place in the universe were fundamentally juridicial. They served to define man's obligations to his fellow men and to nature. The Judaic tradition, for example, rests on the idea of a contractual relationship between God and man. This relationship must and does leave room for autonomy for both God and man, for a contract is an agreement willingly entered into by parties who are free not to agree. Man's autonomy and his corresponding responsibility is a central issue of all religious systems. The spiritual cosmologies engendered by modern science, on the other hand, are infected with the germ of logical necessity. They, except in the hands of the wisest scientists and philosophers, no longer content themselves with explanations of appearances, but claim to say how things actually are and must necessarily be. In short, they convert truth to provability.

As one consequence of this drive of modern science, the question, "What aspects of life are formalizable?" has been transformed from the moral question, "How and in what form may man's obligations and responsibilities be known?" to the question, "Of what technological genus is man a species?" Even some philosophers whose every instinct rebels against the idea that man is entirely comprehensible as a machine have succumbed to this spirit of the times. Hubert Dreyfus, for example, trains the heavy guns of phenomenology on the computer model of man.[7] But he limits his argument to the technical question of what computers can and cannot do. I would argue that if computers could imitate man in every respect—which in fact they cannot—even then it would be appropriate, nay, urgent, to examine the computer in the light of man's perennial need to find his place in the world. The outcomes of practical matters that are of vital importance to everyone hinge on how and in what terms the discussion is carried out.

One position I mean to argue appears deceptively obvious: it is simply that there are important differences between men and machines as thinkers. I would argue that, however intelligent machines may be made to be, there are some acts of thought that *ought* to be attempted only by humans. One socially significant question I thus intend to raise is over the proper place of computers in the social order. But, as we shall see, the issue transcends computers in that it must ultimately deal with logicality itself—quite apart from whether logicality is encoded in computer programs or not.

The lay reader may be forgiven for being more than slightly incredulous that anyone should maintain that human thought is entirely computable. But his very incredulity may itself be a sign of how marvelously subtly and seductively modern science has come to influence man's imaginative construction of reality.

Surely, much of what we today regard as good and useful, as well as much of what we would call knowledge and wisdom, we owe to science. But science may also be seen as an addictive drug. Not only has our unbounded feeding on science caused us to become dependent on it, but, as happens with many other drugs taken in increasing dosages, science has been gradually converted into a slow-acting poison. Beginning perhaps with Francis Bacon's misreading of the genuine promise of science, man has been seduced into wishing and working for the establishment of an age of rationality, but with his vision of rationality tragically twisted so as to equate it with logicality. Thus have we very nearly come to the point where almost every genuine human dilemma is seen as a mere paradox, as a merely apparent contradiction that could be untangled by judicious applications of cold logic derived from a higher standpoint. Even murderous wars have come to be perceived as mere problems to be solved by hordes of professional problemsolvers. As Hannah Arendt said about recent makers and executors of policy in the Pentagon:

"They were not just intelligent, but prided themselves on being 'rational' . . . They were eager to find formulas, preferably expressed in a pseudo-mathematical language, that would unify the most disparate phenomena with which reality presented them; that is, they were eager to discover *laws* by which to explain and predict political and historical facts as though they were as necessary, and thus as reliable, as the physicists once believed natural phenomena to be . . . [They] did not *judge;* they calculated. . . . an utterly irrational confidence in the calculability of reality [became] the leit-motif of the decision making."[8]

And so too have nearly all political confrontations, such as those between races and those between the governed and their governors, come to be perceived as mere failures of communication. Such rips in the social fabric can then be systematically repaired by the expert application of the latest information-handling techniques—at least so it is believed. And so the rationality-is-logicality equation, which the very success of science has drugged us into adopting as virtually an axiom, has led us to deny the very existence of human conflict, hence the very possibility of the collision of genuinely incommensurable human interests and of disparate human values, hence the existence of human values themselves.

It may be that human values are illusory, as indeed B. F. Skinner argues. If they are, then it is presumably up to science to demonstrate that fact, as indeed Skinner (as scientist) attempts to do. But then science must itself be an illusory system. For the only certain knowledge science can give us is knowledge of the behavior of formal systems, that is, systems that are games invented by man himself and in which to assert truth is nothing more or less than to assert that, as in a chess game, a particular board position was arrived at by a sequence of legal moves. When science purports to make statements about man's experiences, it bases them on identifications between the primitive (that is, undefined) objects of one of its formalisms, the pieces of one of its games, and some set of human observations. No such sets of correspondences can ever be proved to be correct. At best, they can be falsified, in the sense that formal manipulations of a system's symbols may lead to symbolic configurations which, when read in the light of the set of correspondences in question, yield interpretations contrary to empirically observed phenomena. Hence all empirical science is an elaborate structure built on piles that are anchored, not on bedrock as is commonly supposed, but on the shifting sand of fallible human judgment, conjecture, and intuition. It is not even true, again contrary to common belief, that a single purported counter-instance that, if accepted as genuine would certainly falsify a specific scientific theory, generally leads to the immediate abandonment of that theory. Probably all scientific theories currently accepted by scientists themselves (excepting only those purely formal theories claiming no relation to the empirical world) are today confronted with contradicting evidence of more than negligible weight that, again if fully credited, would logically invalidate them. Such evidence is often explained (that is, explained away) by ascribing it to error of some kind, say, observational error, or by characterizing it as inessential, or by the assumption (that is, the faith) that some yet-to-be-discovered way of dealing with it will some day permit it to be acknowledged but nevertheless incorporated into the scientific theories it was originally thought to contradict. In this way scientists continue to rely on already impaired theories and to infer "scientific fact" from them.*

The man in the street surely believes such scientific facts to be as well-established, as well-proven, as his own existence. His certitude is an illusion. Nor is the scientist himself immune to the same illusion. In his praxis, he must, after all, suspend disbelief in order to do or think anything at all. He is rather like a theatergoer, who, in order to participate in and understand what is happening on the stage, must for a time pretend to himself that he is witnessing real events. The scientist must believe his working hypothesis, together with its vast underlying structure of theories and assumptions, even if only for the sake of the argument. Often the "argument" extends over his entire lifetime. Gradually he becomes what he at first merely pretended to be: a true believer. I choose the word "argument" thoughtfully, for scientific demonstrations, even mathematical proofs, are fundamentally acts of persuasion.

---

* Thus, Charles Everett writes on the now-discarded phlogiston theory of combustion (in the *Encyclopaedia Britannica*, 11th ed., 1911, vol. VI, p. 34): "The objections of the anti-phlogistonists, such as the fact that the calices weigh more than the original metals instead of less as the theory suggests, were answered by postulating that phlogiston was a principle of levity, or even completely ignored as an accident, the change in qualities being regarded as the only matter of importance." Everett lists H. Cavendish and J. Priestley, both great scientists of their time, as adherents to the phlogiston theory.

Scientific statements can never be certain; they can be only more or less credible. And credibility is a term in individual psychology, i.e., a term that has meaning only with respect to an individual observer. To say that some proposition is credible is, after all, to say that it is believed by an agent who is free not to believe it, that is, by an observer who, after exercising judgment and (possibly) intuition, chooses to accept the proposition as worthy of his believing it. How then can science, which itself surely and ultimately rests on vast arrays of human value judgments, demonstrate that human value judgments are illusory? It cannot do so without forfeiting its own status as the single legitimate path to understanding man and his world.

But no merely logical argument, no matter how cogent or eloquent, can undo this reality: that science has become the sole legitimate form of understanding in the common wisdom. When I say that science has been gradually converted into a slow-acting poison, I mean that the attribution of certainty to scientific knowledge by the common wisdom, an attribution now made so nearly universally that it has become a commonsense dogma, has virtually delegitimatized all other ways of understanding. People viewed the arts, especially literature, as sources of intellectual nourishment and understanding, but today the arts are perceived largely as entertainments. The ancient Greek and Oriental theaters, the Shakespearian stage, the stages peopled by the Ibsens and Chekhovs nearer to our day—these were schools. The curricula they taught were vehicles for understanding the societies they represented. Today, although an occasional Arthur Miller or Edward Albee survives and is permitted to teach on the New York or London stage, the people hunger only for what is represented to them to be scientifically validated knowledge. They seek to satiate themselves at such scientific cafeterias as *Psychology Today,* or on popularized versions of the works of Masters and Johnson, or on scientology as revealed by L. Ron Hubbard. Belief in the rationality-logicality equation has corroded the prophetic power of language itself. We can count, but we are rapidly forgetting how to say what is worth counting and why.

1

ON TOOLS

The stories of man and of his machines are inseparably woven together. Machines have enabled man to transform his physical environment. With their aid he has plowed the land and built cities and dug great canals. These transformations of man's habitat have necessarily induced mutations in his societal arrangements. But even more crucially, the machines of man have strongly determined his very understanding of his world and hence of himself. Man is conscious of himself, of the existence of others like himself, and of a world that is, at least to some extent, malleable. Most importantly, man can foresee. In the act of designing implements to harrow the pliant soil, he rehearses their action in his imagination. Moreover, since he is conscious of himself as a social creature and as one who will inevitably die, he is necessarily a teacher. His tools, whatever their primary practical function, are necessarily also pedagogical instruments. They are then part of the stuff out of which man fashions his imaginative reconstruction of the world. It is within the intellectual and social world he himself creates that the individual prehearses and rehearses countless dramatic enactments of how the world might have been and what it might become. That world is the repository of his subjectivity. Therefore it is the stimulator of his consciousness and finally the constructor of the material world itself. It is this self-constructed world that the individual encounters as an apparently external force. But he contains it within himself; what confronts him is his own model of a universe, and, since he is part of it, his model of himself.

Man can create little without first imagining that he can create it. We can imagine the rehearsal of how he would use it that must have gone on in a stone-age man while he laboriously constructed his axe. Did not each of us recapitulate this ancestral experience when as small children we constructed primitive toys of whatever material lay within our reach? But tools and machines do not merely signify man's imaginativeness and its creative reach, and they are certainly not important merely as instruments for the transformation of a malleable earth: they are pregnant symbols in themselves. They symbolize the activities they enable, i.e., their own use. An oar is a tool for rowing, and it represents the skill of rowing in its whole complexity. No one who has not rowed can see an oar as truly an oar. The way someone who has never played one sees the violin is simply not the same, by very far, as the way a violinist sees it. A tool is also a model for its own reproduction and a script for the reenactment of the skill it symbolizes. That is the sense in which it is a pedagogic instrument, a vehicle for instructing men in other times and places in culturally acquired modes of thought and action. The tool as symbol in all these respects thus transcends its role as a practical means toward certain ends: it is a constituent of man's symbolic recreation of his world. It must therefore inevitably enter into the imaginative calculus that constantly constructs his world. In that sense, then, the tool is much more than a mere device: it is an agent for change. It is even more than a fragment of a blueprint of a world determined for man and bequeathed to him by his forebearers—although it is that too.

It is readily understandable that hand-held tools and especially hand-held weapons have direct effects on the imaginations of individuals who use them. When hunters acquired spears, for example, they must have seen themselves in an entirely new relationship to their world. Large animals which had earlier raided their foodstores and even attacked their children and which they feared, now became man's prey. Man's source of food grew, for now men could kill animals at a distance, including many species that had eluded them before. The effectively greater abundance of food must also have enlarged the domain over which they could range, thus increasing the likelihood that they would meet other people. Their experience of the world changed and so too must have their idea of their place in it.

The six-shooter of the nineteenth-century American West was known as the "great equalizer," a name that eloquently testifies to what that piece of hardware did to the self-image of gun-toters who, when denuded of their weapons, felt themselves disadvantaged with respect to their fellow citizens. But devices and machines, perhaps known to (and certainly owned and operated by) only a relatively few members of society, have often influenced the self-image of its individual members and the world-image of the society as a whole quite as profoundly as have widely used hand tools. Ships of all kinds, for example, were instrumental in informing man of the vastness of his domain. They permitted different cultures to meet and to cross-fertilize one another. The seafarer's ships and all his other artifacts, his myths and legends, effectively transmitted his lore from generation to generation. And they informed the unconscious of those who stayed on the land as much as that of those who actually sailed. The printing press transformed the world even for those millions who, say, in Martin Luther's time, remained illiterate and perhaps never actually saw a book. And of the great masses of people all over the world whose lives were directly and dramatically changed by the industrial revolution, how many ever actually operated a steam engine? Nor is modern society immune to huge shocks administered as side effects of the introduction of new machines. The cotton-picking machine was deployed in the cotton fields of the American South beginning about 1955. It quickly destroyed the market for the only thing vast masses of black Southern agricultural workers had to sell: their labor. Thus began the mass migration of the American Black to the cities, particularly to such northern manufacturing centers as Detroit, Chicago, and New York, but also to the large Southern cities, such as Birmingham and Atlanta. Surely this enormous change in the demography of the United States, this internal migration of millions of its citizens, was and remains one of the principal determinants of the course of the American civil-rights movement. And that movement has nontrivially influenced the consciousness of every American at least, if not of almost every living adult anywhere on this earth.

What is the compelling urgency of the machine that it can so intrude itself into the very stuff out of which man builds his world?

Many machines are functional additions to the human body, virtually prostheses. Some, like the lever and the steam shovel extend the raw muscular power of their individual operators; some, like the microscope, the telescope, and various measuring instruments, are extensions of man's sensory apparatus. Others extend the physical reach of man. The spear and the radio, for example, permit man to cast his influence over a range exceeding that of his arms and voice, respectively. Man's vehicles make it possible for him to travel faster and farther than his legs alone would carry him, and they allow him to transport great loads over vast distances. It is easy to see how and why such prosthetic machines directly enhance man's sense of power over the material world. And they have an important psychological effect as well: they tell man that he can remake himself. Indeed, they are part of the set of symbols man uses to recreate his past, i.e., to construct his history, and to create his future. They signify that man, the engineer, can transcend limitations imposed on him by the puniness of his body and of his senses. Once man could kill another animal only by crushing or tearing it with his hands; then he acquired the axe, the spear, the arrow, the ball fired from a gun, the explosive shell. Now charges mounted on missiles can destroy mankind itself. That is one measure of how far man has extended and remade himself since he began to make tools.

To construe the influence of prosthetic tools on man's transformation entirely in terms of the power they permitted man to aggregate to himself may invite a view of man's relationship to nature whose principal—indeed, almost sole—component is a raw power struggle. Man, in this view, finally conquered nature simply by mustering sufficient power to overcome natural space and time, to engineer life and death, and finally to destroy the earth altogether. But this idea is mistaken, even if we accept that man's eternal dream has been, not merely the discovery of nature, but its conquest, and that that dream has now been largely realized. For if victory over nature has been achieved in this age, then the nature over which modern man reigns is a very different nature from that in which man lived before the scientific revolution. Indeed, the trick that man turned and that enabled the rise of modern science was nothing less than the transformation of nature and of man's perception of reality.

The paramount change that took place in the mental life of man, beginning during roughly the fourteenth century, was in man's perception of time and consequently of space. Man had long ago noticed (and, we may suppose, thought about) regularities in the world about him. Alexander Marshack has shown that even Upper Paleolithic man (circa 30,000 B.C.) had a notation for lunar cycles that was, in Marshack's words, "already evolved, complex and sophisticated, a tradition that would seem to have been thousands of years old by this point."[1] But from Classical antiquity until relatively recently, the regularity of the universe was searched for and perceived in great thematic harmonies. The idea that nature behaves systematically in the sense we understand it—i.e., that every part and aspect of nature may be isolated as a subsystem governed by laws describable as functions of time—this idea could not have been even understood by people who perceived time, not as a collection of abstract units (i.e., hours, minutes, and seconds), but as a sequence of constantly recurring events.

Times of day were known by events, such as the sun standing above a specific pile of rocks, or, as Homer tells us, by tasks begun or ended, such as the yoking of the oxen (morning) and the unyoking of the oxen (evening). Durations were indicated by reference to common tasks, e.g., the time needed to travel a well-known distance or to boil fixed quantities of water. Seasonal times were known by recurring seasonal events, e.g., the departure of birds.

Until Darwin's theory of evolution began to sink into the stream of commonly held ideas, i.e., to become "common sense," people knew that the world about them—the world of reproducing plants and animals, of rivers that flowed and dried up and flowed again, of seas that pulsed in great tidal rhythms, and of the ever-repeating spectacles in the heavens—had always existed, and that its fundamental law was eternal periodicity. Cosmological time, as well as the time perceived in daily life, was therefore a sort of complex beating, a repeating and echoing of events. Perhaps we can vaguely understand it by contemplating, say, the great fugues of Bach. But a special form of contemplation is required of us: we must not think in the modern manner, i.e., of Bach as a "problem solver," or of each of his *opera* in his *Art of the Fugue* as being his increasingly refined "solution" to a problem he had originally set himself. Instead we must think that Bach had the whole plan in his mind all the time, that he thought of the *Art of the Fugue* as a unified work with no beginning and no end, itself eternal like the cosmos, and like it enormously intricate in its connections, circles within circles within circles. We might then find it possible to think of life as having been not merely punctuated but entirely suffused by this kind of music, both on the grand cosmological-theological scale and on the small day-to-day level. Such time is a revolution of cycles and epicycles within cycles, not the receptacle of a uniformly flowing progression of abstract moments we now "know" it to be. And nature itself consisted, to be sure, of individual phenomena, but individual phenomena that were constantly repeating metamorphoses of themselves, and hence were permanent, eternal. "What is eternal is circular, and what is circular is eternal," Aristotle said, and even Galileo still believed the universe to be eternal and to be governed by recurrence and periodicity.

Darwin's understanding of time was radically different. He saw nature itself as a process in time and the individual phenomena of nature as irreversible metamorphoses. But he was far from being the originator of the idea of progress that is now so much with us. Indeed, he would not have been able to think his thoughts at all, if something very nearly like our current ideas of time had not already been an integral part of the common sense of his era.

How man's perception of time changed from that of the ancients to ours illuminates the role played by another kind of machine (one that is not prosthetic) in man's transformation from a creature of and living in nature to nature's master.

The clock is not a prosthetic machine; its product is not an extension of man's muscles or senses, but hours, minutes, and seconds, and today even micro-, nano-, and pico-seconds. Lewis Mumford calls the clock, not the steam engine, "the key machine of the modern industrial age."[2] In the brilliant opening chapter of his *Technics and Civilization,* he describes, among other things, how during the Middle Ages the ordered life of the monasteries affected life in the communities adjacent to them.

> "The monastery was the seat of a regular life. . . . the habit of order itself and the earnest regulation of time-sequences had become almost second nature in the monastery. . . . the monasteries—at one time there were 40,000 under the Benedictine rule— helped to give human enterprise the regular collective beat and rhythm of the machine; for the clock is not merely a means of keeping track of the hours, but of synchronizing the actions of men. . . . by the thirteenth century there are definite records of mechanical clocks, and by 1370 a well-designed 'modern' clock had been built by Heinrich von Wyck at Paris. Meanwhile, bell towers had come into existence, and the new clocks, if they did not have, till the fourteenth century, a dial and a hand that translated the movement of time into a movement through space, at all events struck the hours. The clouds that could paralyze the sundial . . . were no longer obstacles to time-keeping: summer or winter, day or night, one was aware of the measured clank of the clock. The instrument presently spread outside the monastery; and the regular striking of the bells brought a new regularity into the life of the workman and the merchant. The bells of the clock tower almost defined urban existence. Time-keeping passed into time-serving and time-accounting and time-rationing. As this took place, Eternity ceased gradually to serve as the measure and focus of human actions."[3]

Mumford goes on to make the crucial observation that the clock "disassociated time from human events and helped create the belief in an independent world of mathematically measurable sequences: the special world of science."[4] The importance of that effect of the clock on man's perception of the world can hardly be exaggerated. Our current view of time is so deeply ingrained in us, so much "second nature" to us, that we are virtually incapable any longer of identifying the role it plays in our thinking. Alexander Marshack remarks:

> "The concept of the time-factored process in the hard sciences is today almost tautological, since all processes, simple or complex, sequential or interrelated, finite or infinite, develop or continue and have measurable or estimable rates, velocities, durations, periodicities, and so on. However, the sciences which study these processes are themselves 'time-factored,' since the processes of cognition and recognition, of planning, research, analysis, comparison, and interpretation are also sequential, interrelated, developmental and cumulative."[5]

Indeed, the two fundamental equations of physics that every high-school student knows are $F = ma$ and $E = mc^2$. The a in the first stands for acceleration, i.e., a change of velocity with time, and the c in the second stands for the velocity of light, i.e., the displacement of light with time.

I mention the clock here not merely because it was a crucial determinant of man's thinking—there were, after all, many other inventions that helped initiate the new scientific rationalism; for example, lines of longitude and latitude on the globe—but to show that prosthetic machines alone do not account for man's gain of power over nature. The clock is clearly not a prosthetic machine; it extends neither man's muscle power nor his senses. It is an autonomous machine.

Many machines are automatic in the sense that, once they are turned on, they may run by themselves for long periods of time. But most automatic machines have to be set to their task and subsequently steered or regulated by sensors or by human drivers. An autonomous machine is one that, once started, runs by itself on the basis of an internalized model of some aspect of the real world. Clocks are fundamentally models of the planetary system. They are the first autonomous machines built by man, and until the advent of the computer they remained the only truly important ones.

Where the clock was used to reckon time, man's regulation of his daily life was no longer based exclusively on, say, the sun's position over certain rocks or the crowing of a cock, but was now based on the state of an autonomously behaving model of a phenomenon of nature. The various states of this model were given names and thus reified. And the whole collection of them superimposed itself on the existing world and changed it, just as much as a cataclysmic rearrangement of its geography or climate might have changed it. Man now had to develop new senses for finding his way about the world. The clock had created literally a new reality; and that is what I meant when I said earlier that the trick man turned that prepared the scene for the rise of modern science was nothing less than the transformation of nature and of his perception of reality. It is important to realize that this newly created reality was and remains an impoverished version of the older one, for it rests on a rejection of those direct experiences that formed the basis for, and indeed constituted, the old reality. The feeling of hunger was rejected as a stimulus for eating; instead, one ate when an abstract model had achieved a certain state, i.e., when the hands of a clock pointed to certain marks on the clock's face (the anthropomorphism here is highly significant too), and similarly for signals for sleep and rising, and so on.

This rejection of direct experience was to become one of the principal characteristics of modern science. It was imprinted on western European culture not only by the clock but also by the many prosthetic sensing instruments, especially those that reported on the phenomena they were set to monitor by means of pointers whose positions were ultimately translated into numbers. Gradually at first, then ever more rapidly and, it is fair to say, ever more compulsively, experiences of reality had to be representable as numbers in order to appear legitimate in the eyes of the common wisdom. Today enormously intricate manipulations of often huge sets of numbers are thought capable of producing new aspects of reality. These are validated by comparing the newly derived numbers with pointer readings on still more instruments that mediate between man and nature, and which, of course, produce still more numbers.

"The scientific man has above all things to strive at self-elimination in his judgments," wrote Karl Pearson in 1892.[6] Of the many scientists I know, only a very few would disagree with that statement. Yet it must be acknowledged that it urges man to strive to become a disembodied intelligence, to himself become an instrument, a machine. So far has man's initially so innocent liaison with prostheses and pointer readings brought him. And upon a culture so fashioned burst the computer.

"Every thinker," John Dewey wrote, "puts some portion of an apparently stable world in peril and no one can predict what will emerge in its place." So too does everyone who invents a new tool or, what amounts to the same thing, finds a new use for an old one. The long historical perspective which aids our understanding of Classical antiquity, of the Middle Ages, and of the beginnings of the Modern Age also helps us to formulate plausible hypotheses to account for the new realities which emerged in those times to replace older ones imperiled by the introduction of new tools. But as we approach the task of understanding the warp and woof of the stories that tell, on the one hand, of the changing consciousness of modern man, and, on the other, of the development of contemporary tools and particularly of the computer, our perspective necessarily flattens out. We have little choice but to project the lessons yielded by our understanding of the past, our plausible hypotheses, onto the present and the future. And the difficulty of that task is vastly increased by the fact that modern tools impact on society far more critically in a much shorter time than earlier ones did.

The impulse the clock contributed toward the allienation of man from nature required centuries to penetrate and decisively affect mankind as a whole. And even then, it had to synergistically combine with many other emerging factors to exercise its influence. The steam engine arrived when, in the common-sense view, time and space were already quantified. An eternal nature governed by immutable laws of periodicity implied a mandate, one made explicit in holy books and exercised by institutional vicars of the eternal order. That quasi-constitutional, hence constrained, authority had long since been displaced by, for example, the relatively unconstrained authority of money, i.e., of value quantified, and especially the value of a man's labor quantified. These and many other circumstances combined to make it possible for the steam engine to eventually transform society radically. Later tools, e.g., the telephone, the automobile, radio, impinged on a culture already enthralled by what economists call the pig principle: if something is good, more is better. The hunger for more communication capacity and more speed, often stimulated by the new devices themselves, as well by new marketing techniques associated with them, enabled their rapid spread throughout society and society's increasingly rapid transformation under their influence.

When the first telegraph line connecting Texas with New York was laid, doubts were expressed as to whether the people in those places would have anything to say to one another. But by the time the digital computer emerged from university laboratories and entered the American business, military, and industrial establishments, there were no doubts about its potential utility. To the contrary, American managers and technicians agreed that the computer had come along just in time to avert catastrophic crises: were it not for the timely introduction of computers, it was argued, not enough people could have been found to staff the banks, the ever increasingly complex communication and logistic problems of American armed forces spread all over the world could not have been met, and trading on the stock and commodity exchanges could not have been maintained. The American corporation was faced with a "command and control" problem similar to that confronting its military counterpart. And like the Pentagon, it too was increasingly diversified and internationalized. Unprecedentedly large and complex computational tasks awaited American society at the end of the Second World War, and the computer, almost miraculously it would seem, arrived just in time to handle them.

In fact, huge managerial, technological, and scientific problems had been solved without the aid of electronic computers in the decades preceding the Second World War and especially during the war itself. A dominant fraction of the industrial plant of the United States was coordinated to provide the tools of war—foodstuffs, clothing, etc.—and to supply the required transport to vast armies spread all over the globe. The Manhattan Project produced the atomic bomb without using electronic computers; yet the scientific and engineering problems solved under its auspices required probably more computations than had been needed for all astronomical calculations performed up to that time. The magnitude of its managerial task surely rivaled that of the Apollo Project of the sixties. Most people today probably believe that the Apollo Project could not have been managed without computers. The history of the Manhattan Project seems to contradict that belief. There are corresponding beliefs about the need for computers in the management of large corporations and of the military, about the indispensability of computers in modern scientific computations, and, indeed, about the impossibility of pursuing modern science and modern commerce at all without the aid of computers.*

The belief in the indispensability of the computer is not entirely mistaken. The computer becomes an indispensable component of any structure once it is so thoroughly integrated with the structure, so enmeshed in various vital substructures, that it can no longer be factored out without fatally impairing the whole structure. That is virtually a tautology. The utility of this tautology is that it can reawaken us to the possibility that some human actions, e.g., the introduction of computers into some complex human activities, may constitute an irreversible commitment. It is not true that the American banking system or the stock and commodity markets or the great manufacturing enterprises would have collapsed had the computer not come along "just in time." It is true that the specific way in which these systems actually developed in the past two decades, and are still developing, would have been impossible without the computer. It is true that, were all computers to suddenly disappear, much of the modern industrialized and militarized world would be thrown into great confusion and possibly utter chaos. The computer was not a prerequisite to the survival of modern society in the postwar period and beyond; its enthusiastic, uncritical embrace by the most "progressive" elements of American government, business, and industry quickly made it a resource essential to society's sur-

---

* I am sure that, had computers attained their present sophistication by 1940, technicians participating in the Manhattan Project would have sworn that it too would have been impossible without computers. And we would have had similarly fervent testimony from the designers of Second World War aircraft, and from the managers of logistics of that war. If Germany had had computers from the outset of Hitler's dictatorship, common sense would today hold that only with the aid of computers could the Nazis have controlled the German people and implemented the systematic transportation of millions of people to death camps and their subsequent murder. But the Second World War was fought, and the millions did die, when there were still no computers.

vival *in the form* that the computer itself had been instrumental in shaping.

In 1947 J. W. Forrester wrote a memorandum to the U.S. Navy "On the Use of Electronic Digital Computers as Automatic Combat Information Centers." Commenting on subsequent developments in 1961, he wrote,

> "one could probably not have found [in 1947] five military officers who would have acknowledged the possibility of a machine's being able to analyze the available information sources, the proper assignment of weapons, the generation of command instructions, and the coordination of adjacent areas of military operations. . . . During the following decade the speed of military operations increased until it became clear that, regardless of the assumed advantages of human judgment decisions, the internal communication speed of the human organization simply was not able to cope with the pace of modern air warfare. This inability to act provided the incentive."[7]

The decade of which Forrester speaks was filled with such incentives, with discoveries that existing human organizations were approaching certain limits to their ability to cope with the ever faster pace of modern life. The image Forrester invokes is of small teams of men hurrying to keep up with events but falling ever further behind because things are happening too fast and there is too much to do. They have reached the limit of the team's "internal speed." Perhaps this same imagery may serve as a provocative characterization also for teams of bank clerks frantically sorting and posting checks in the middle of the night, attacking ever larger mountains of checks that must, according to law, be cleared by a fixed deadline. Perhaps all, or at least many, of the limits of other kinds that were being approached during that decade may usefully be so characterized. After all, it is ultimately the "internal speed" of some human organization that will prove the limiting factor when, say, an automobile firm attempts to run a production line capable of producing an astronomical variety of cars at a high and constant rate, or when, say, some central government agency takes the responsibility for guarding millions of welfare clients against the temptation to cheat by closely monitoring both their welfare payments and whatever other income they may, possibly illicitly, receive.

The "inability to act" which, as Forrester points out, "provided the incentive" to augment or replace the low-internal-speed human organizations with computers, might in some other historical situation have been an incentive for modifying the task to be accomplished, perhaps doing away with it altogether, or for restructuring the human organizations whose inherent limitations were, after all, seen as the root of the trouble. It may be that the incentive provided by the military's inability to cope with the increasing complexity of air warfare in the 1950's could have been translated into a concern, not for mustering techniques to enable the military to keep up with their traditional missions, but for inventing new human organizations with new missions, missions relevant to more fundamental questions about how peoples of diverse interests are to live with one another. But the computer was used to build, in the words of one air force colonel, "a servomechanism spread out over an area comparable to the whole American continent," that is, the SAGE air-defense system. Of course, once "we" had such a system, we had to assume "they" had one too. We therefore had to apply our technology to designing offensive weapons and strategies that could overpower "our" defenses, i.e., "their" presumed defenses. We then had to assume that "they" had similar weapons and strategies and therefore . . ., and so on to today's MIRVs and MARVs and ABMs.

It may be that the people's cultivated and finally addictive hunger for private automobiles could have been satiated by giving them a choice among, say, a hundred vehicles that actually differ substantially from one another, instead of a choice among the astronomical number of basically identical "models" that differ only trivially from one another. Indeed, perhaps the private automobile could have been downgraded as a means of personal transportation in favor of mass transit in, and passenger rail between, the cities. But the computer was used to automate the flow of parts to automobile production lines so that people could choose from among millions of trivial options on their new cars.

It may be that social services such as welfare could have been administered by humans exercising human judgment if the dispensing of such services were organized around decentralized, indigenous population groupings, such as neighborhoods and natural regions. But the computer was used to automate the administration of social services and to centralize it along established political lines. If the computer had not facilitated the perpetuation and "improvement" of existing welfare distribution systems—hence of their philosophical rationales—perhaps someone might have thought of eliminating much of the need for welfare by, for example, introducing negative income tax. The very erection of an enormously large and complex computer based welfare administration apparatus, however, created an interest in its maintenance and therefore in the perpetuation of the welfare system itself. And such interests soon become substantial barriers to innovation even if good reasons to innovate later accumulate. In other words, many of the problems of growth and complexity that pressed insistently and irresistibly for response during the postwar decades could have served as incentives for social and political innovation. An enormous acceleration of social invention, had it begun then, would now seem to us as natural a consequence of man's predicament in that time as does the flood of technological invention and innovation that was actually stimulated.

Yes, the computer did arrive "just in time." But in time for what? In time to save—and save very nearly intact, indeed, to entrench and stabilize—social and political structures that otherwise might have been either radically renovated or allowed to totter under the demands that were sure to be made on them. The computer, then, was used to conserve America's social and political institutions. It buttressed them and immunized them, at least temporarily, against enormous pressures for change. Its influence has been substantially the same in other societies that have allowed the computer to make substantial inroads upon their institutions: Japan and Germany immediately come to mind.

The invention of the computer put a portion of an apparently stable world in peril, as it is the function of almost every one of man's creative acts to do. And, true to Dewey's dictum, no one could have predicted what would emerge in its place. But of the many paths to social innovation it opened to man, the most fateful was to make it possible for him to eschew all deliberate thought of substantive change. That was the option man chose to exercise. The arrival of the Computer Revolution and the founding of the Computer Age have been announced many times. But if the triumph of a revolution is to be measured in terms of the profundity of the social revisions it entrained, then there has been no computer revolution. And however the present age is to be characterized, the computer is not eponymic of it.

To say that the computer was initially used mainly to do things pretty much as they had always been done, except to do them more rapidly or, by some criteria, more efficiently, is not to distinguish it from other tools. Only rarely, if indeed ever, are a tool and an altogether original job it is to do, invented together. Tools as symbols, however, invite their imaginative displacements into other than their original contexts. In their new frames of reference, that is, as new symbols in an already established imaginative calculus, they may themselves be transformed, and may even transform the originally prescriptive calculus. These transformations may, in turn, create entirely new problems which then engender the invention of hitherto literally unimaginable tools. In 1804, a hundred years after the first stationary steam engines of Newcomen and Savery had found common use in England to, for example, pump water out of mines, Trevithik put a steam engine on a carriage and the carriage on the tracks of a horse-tramway in Wales. This ripping out of context of the stationary steam engine and its displacement into an entirely new context transformed the engine into a locomotive, and began the transformation of the horse-tramway into the modern railroad. And incidentally, since it soon became necessary to guard against collisions of trains traveling on the same track, a whole new signaling technology was stimulated. New problems had been created and, in response to them, new tools invented.

It is noteworthy that Thomas Savery, the builder of the first steam engine that was applied practically in industry (circa 1700), was also the first to use the term "horsepower" in approximately its modern sense. Perhaps the term arose only because there were so many horses when the steam engine replaced them, not only in its first incarnation as a stationary power source, but also in its reincarnation as a locomotive. Still, the term "horsepower," so very pointed in its suggestiveness, might well have provoked Trevithik's imagination to probe in the direction it finally moved, to make the creative leap that combined the steam engine and the horse-tramway in a single unified frame of reference. Invention involves the imaginative projection of symbols from one existing, and generally well-developed, frame of reference to another. It is to be expected that some potent symbols will survive the passage nearly intact, and will exert their influence on even the new framework.

Computers had horses of another color to replace. Before the first modern electronic digital computers became available for what we now call business data processing—that is, before the acquisition of UNIVAC I by the U.S. Bureau of the Census in 1951—many American businesses operated large so-called "tab rooms." These rooms housed machines that could punch the same kind of cards (now commonly, if often mistakenly, called IBM cards) that are still in use today, sort these cards according to arbitrary criteria, and "tabulate" decks of such cards, i.e., list the information they contained in long printed tables. Tab rooms produced mountains of management reports for American government and industry, using acres of huge clanking mechanical monsters. These machines could perform only one operation on a deck of cards at a time. They could, for example, sort the deck on a specific sorting key. If the sorted deck had to be further sorted according to yet another criterion, the new criterion had to be manually set into the machine and the deck fed through the machine once more. Tab rooms were the horse-tramways of business data processing, tab machines the horses.

In principle, even the earliest commercially available electronic computers, the UNIVAC I's, made entirely new and much more efficient data-processing techniques possible, just as, in principle, the earliest steam engines could already have been mounted on carriages and the carriages on tracks. Indeed, during and just after the Second World War, the arts of operations research and systems analysis, on which the sophisticated use of computers in business was ultimately grounded, were developed to very nearly their full maturity. Still, business used the early computers to simply "automate" its tab rooms, i.e., to perform exactly the earlier operations, only now automatically and, presumably, more efficiently. The crucial transition, from the business computer as a mere substitute for work-horse tab machines to its present status as a versatile information engine, began when the power of the computer was projected onto the framework already established by operations research and systems analysis.

It must be added here that although the railroad in England became important in its own right—it employed many workers, for example—it also enormously increased the importance of many other forms of transportation. Similarly, the synergistic combination of computers and systems analysis played a crucial role in the creation and growth of the computer industry. It also breathed a new vitality into systems analysis as such. During the first decade of the computer's serious invasion of business, when managers often decided their businesses needed computers even though they had only the flimsiest bases for such decisions, they also often undertook fairly penetrating systems analyses of their operations in order to determine what their new computers were to do. In a great many cases such studies revealed opportunities to improve operations, sometimes radically, without introducing computers at all. Nor were computers used in the studies themselves. Often, of course, computers were installed anyway for reasons of, say, fashion or prestige.

A side effect of this oft-repeated experience was to firmly establish systems analysis, and to a lesser extent operations research, as a methodology for making business decisions. As the prestige of systems analysis was fortified by its successes and as, simultaneously, the computer grew in power, the problems tackled by systems analysts became more and more complex, and the computer appeared an ever more suitable instrument to handle great complexity. Normally systems analysis appears, to the casual observer at least, to have been swallowed up by the computer. This appearance is misleading but not without significance. Systems analysis has survived and prospered as a discipline in its own right. The computer has put muscles on its techniques. It has so greatly strengthened them as to make them qualitatively different from their early manual counterparts. The latter, consequently, have largely disappeared. And the computer can no longer be factored out of the former.

The interaction of the computer with systems analysis is instructive from another point of view as well. It is important to understand very clearly that strengthening a particular technique—putting muscles on it—contributes nothing to its validity. For example, there are computer programs that carry out with great precision all the calculations required to cast the horoscope of an individual whose time and place of birth are known. Because the computer does all the tedious symbol manipulations, they can be done much more quickly and in much more detail than is normally possible for a human astrologer. But such an improvement in the technique of horoscope casting is irrelevant to the validity of astrological forecasting. If astrology is nonsense, then computerized astrology is just as surely nonsense. Now, sometimes certain simple techniques are invalid for the domains to which they are applied merely because of their very simplicity, whereas much more complicated techniques of the *same* kind are valid in those domains. That is not true for astrology, but may well be true of, say, numerical weather forecasting. For the latter, the number of data that must be taken into account, and the amount of computation that must be done on them in order to produce an accurate weather forecast, may well be so large that no team of humans, however large, could complete the computations in any reasonable time whatever. And any simplification of the technique sufficient to reduce the computational task to proportions manageable by humans would invalidate the technique itself. In such cases the computer may contribute to making a hitherto impractical technique practical. But what has to be remembered is that the validity of a technique is a question that involves the technique and its subject matter. If a bad idea is to be converted into a good one, the *source* of its weakness must be discovered and repaired. A person falling into a manhole is rarely helped by making it possible for him to fall faster or more efficiently.

It may seem odd, even paradoxical, that the enhancement of a technique may expose its weaknesses and limitations, but it should not surprise us. The capacity of the human mind for sloppy thinking and for rationalizing, for explaining away the consequences of its sloppy thinking, is very large. If a particular technique requires an enormous amount of computation and if only a limited computational effort can be devoted to it, then a failure of the technique can easily be explained away on the ground that, because of computational limitations, it was never really tested. The technique itself is immunized against critical examination by such evasions. Indeed, it may well be fortified, for the belief that an otherwise faultless and probably enormously powerful technique is cramped by some single limitation tends to lead the devotee to put effort into removing that limitation. When this limitation seems to him to be entirely computational, and when a computer is offered to help remove it, he may well launch a program of intensive, time-consuming "research" aimed simply at "computerizing" his technique. Such programs usually generate subproblems of a strictly computational nature that tend, by virtue of their very magnitude, to increasingly dominate the task and, unless great care is taken to avoid it, to eventually become the center of attention. As ever more investment is made in attacking these initially ancillary subproblems, and as progress is made in cracking them, an illusion tends to grow that real work is being done on the main problem. The poverty of the technique, if it is indeed impotent to deal with its presumed subject matter, is thus hidden behind a mountain of effort, much of which may well be successful in its own terms. But these are terms in a constructed context that has no substantive overlap with, or even relationship to, the context determined by the problem to which the original technique is to be applied. The collection of subproblems together with the lore, jargon, and subtechniques which crystalized around them, becomes reified. The larger this collection is, and the more human energy has been invested in its creation, the more real it seems. And the harder the subproblems were to solve and the more technical success was gained in solving them, the more is the original technique fortified.

I have discussed the role that tools play in man's imaginative reconstruction of his world and in the sharpening of his techniques. However, tools play another related role as well: they constitute a kind of language for the society that employs them, a language of social action. Later on I will say more about language. Let it suffice for now to characterize language somewhat incompletely as consisting of a vocabulary—the words of the language—and a set of rules that determine how individual vocabulary items may be concatenated to form meaningful sentences. I leave to one side for the moment the innumerable mysteries that surround the concept of meaning. I restrict myself to its narrowest conception, namely, that of the action which a particular "sentence" in the language of tools initiates and accomplishes.

Ordinary language gains its expressive power in part from the fact that each of its words has a restricted domain of meaning. It would be impossible to say anything in a language that consisted entirely of pronouns, for example. A tool too gains its power from the fact that it permits certain actions and not others. For example, a hammer has to be rigid. It can therefore not be used as a rope. There can be no such things as general-purpose tools, just as there can be no general-purpose words. We know that the use of specific words in vastly general ways, for example, such words as "like" and "y'know," impoverishes rather than enriches current American English.

Perhaps it is as difficult to invent truly new tools as it is to invent truly new words. But the twentieth century has witnessed the invention of at least a modest number of tools that do actually extend the range of action of which the society is capable. The automobile and the highway, radio and television, and modern drugs and surgical procedures immediately come to mind. These things have enabled society to articulate patterns of action that were never before possible. What is less often said, however, is that the society's newly created ways to act often eliminate the very possibility of acting in older ways. An analogous thing happens in ordinary language. For example, now that the word "inoperative" has been used by high government officials as a euphemism for the word "lie," it can no longer be used to communicate its earlier meaning. Terms like "free" (as in "the free world"), "final solution," "defense," and "aggression" have been so thoroughly debased by corrupt usage that they have become essentially useless for ordinary discourse. Similarly, a highway permits people to travel between the geographical centers it connects, but, because of the side effects that it and other factors synergistically engender, it imprisons poor people in inner cities as effectively as if the cities were walled in. The mass-communication media are sometimes said to have reduced the earth to a global village and to have enabled national and even global town meetings. But, in contrast to the traditional New England town meeting which was—and remains so in my home town—an exercise in *participatory* politics, the mass media permit essentially no talking back. Like highways and automobiles, they enable the society to articulate entirely new forms of social action, but at the same time they irreversibly disable formerly available modes of social behavior.

The computer is, in a sense, a tool of this kind. It helped pry open the door to outer space, and it saved certain societal institutions that were threatened with collapse under the weight of a rapidly growing population. But its impact has also closed certain doors that were once open . . . whether irreversibly or not, we cannot say with certainty. There is a myth that computers are today making important decisions of the kind that were earlier made by people. Perhaps there are isolated examples of that here and there in our society. But the widely believed picture of managers typing questions of the form "What shall we do now?" into their computers and then waiting for their computers to "decide" is largely wrong. What is happening instead is that people have turned the processing of information on which decisions must be based over to enormously complex computer systems. They have, with few exceptions, reserved for themselves the right to make decisions based on the outcome of such computing processes. People are thus able to maintain the illusion, and it is often just that, that they are after all the decisionmakers. But, as we shall argue, a computing system that permits the asking of only certain kinds of questions, that accepts only certain kinds of "data," and that cannot even in principle be understood by those who rely on it, such a computing system has effectively closed many doors that were open before it was installed.

In order to understand how the computer attained so very much power, both as an actor and as a force on the human imagination, we must first discuss where the power of the computer comes from and how the computer does what it does. That is what we shall turn our attention to in the next two chapters.

2

WHERE THE POWER
OF THE COMPUTER
COMES FROM

Were we to see something very strange to us, say, a cloud with straight, sharp edges, we would want to know what it was. And were we told it was a fuba, then we would ask what a fuba was. But there are things all around us that are so constantly part of our lives that they are not strange to us and we don't ask what they are. So it is with machines. The word "machine" calls up images of complex and yet somehow regular motion. The back-and-forth movement of the needle of a sewing machine, so analogous both to the hustle of the gyrating, thrusting connecting rods that drive the locomotive's wheels and to the tremor of the pulsating escapement mechanism of the most delicate watch, such images almost sum up what we mean

---

Chapters 2 and 3 are somewhat technical. The reader who is not comfortable with technical material might either skim these two chapters or postpone reading them until after the rest of the book has been read.

by "machine." Almost. Sufficiently so that we need ask no further what a machine is. Regularity, complexity, motion, power. Still, there is more, and we know it.

We set a punch press into motion, and it mangles the hand of a worker who gets too close to it. The very regularity of the machine is its most fearsome property. We put it to its task and it performs, regularly to be sure, but blindly as well. When we say that justice is blind, we mean to commend it as being almost a machine that performs its function without regard to irrelevant facts—but facts nonetheless. To blind justice, whether the prisoner before the bar is rich or is poor or is a man or is a woman is irrelevant. To the punch press, whether the material in its jaws is a piece of metal or a worker's hand is irrelevant. Like all machines, blind justice and punch presses do only what they are made to do—and that they do exactly.

Machines, when they operate properly, are not merely law abiding; they are embodiments of law. To say that a specific machine is "operating properly" is to assert that it is an embodiment of a law we know and wish to apply. We expect an ordinary desk calculator, for example, to be an embodiment of the laws of arithmetic we all know. Should it deliver what we believe to be a wrong result, our faith in the lawfulness of the machine is so strong that we usually assume we have made an error in punching in our data. It is only when it repeatedly malfunctions that we decide there is "something wrong with the machine." We never believe that the laws of arithmetic have been repealed or amended. But neither do we ever believe that the machine is behaving capriciously, i.e., in an unlawful manner. No, in order to restore it to its proper function we seek to understand why it behaves as it *now* does, i.e., of what law it is now an embodiment. We are pleased when we find, say, a broken gear that *accounts* for its aberrant behavior. We have then discovered its law. We now understand the machine we actually have and are therefore in a position to repair it, i.e., to convert it to the machine we had originally, to an embodiment of the ordinary laws of arithmetic. Indeed, we are often quite distressed when a repairman returns a machine to us with the words, "I don't know what was wrong with it. I just jiggled it, and now it's working fine." He has confessed that he failed to come to understand the law of the broken machine and we infer that he cannot now know, and neither can we or anyone, the law of the "repaired" machine. If we depend on that machine, we have become servants of a law we cannot know, hence of a capricious law. And that is the source of our distress.

The machines that populate our world are no longer exclusively, or even mainly, clanking monsters, the noisy motion of whose parts defines them as machines. We have watches whose works are patterns etched on tiny plastic chips, watches without any moving parts whatever. Even their hands are gone. They tell the time, when commanded to, by displaying illuminated numbers on their faces. The rotating mills that once distributed electrical charges to the spark plugs of our automotive engines have been replaced by small black boxes again containing patterns etched on plastic chips, that silently and motionlessly dole out the required pulses. We call these, and a thousand other devices like them, machines too.

This stretching of the connotative range of the word "machine" has two quite separable significances: First, it testifies that the folk wisdom recognizes the essential characteristic of the machine to be its relentless regularity, its blind obedience of the law of which it is an embodiment. And that regularity, as the folk wisdom perceives correctly too, has little to do with material motion. This is the insight which permits people to talk of, say, a bureaucracy or a system of justice as a machine. Second, it reveals an implicit, though very vague, understanding in the folk wisdom of the idea that one aspect of mechanism has to do with information transfer and not with the transmission of material power. The arrival of all sorts of electronic machines, especially of the electronic computer, has changed our image of the machine from that of a transducer and transmitter of *power* to that of a transformer of *information*.

Many other machines have internal components whose functions are primarily to transmit information, even though the over-all function of these machines is to provide mechanical power. Consider, for example, an ordinary four-cycle gasoline engine. It is, of course, a power generator. One of its components is a tappet rod, a straight steel rod whose bottom end rides on a camshaft and whose top end can lift the exhaust valve of the cylinder to which it belongs.

As the engine turns the main drive shaft, it also turns the camshaft and the cam on which the tappet rod rides. The tappet rod therefore performs an up-down motion which successively and at just the correct times opens and closes the cylinder's exhaust valve. In simple gasoline engines the tappet rod provides both the power to move the valve and the required timing. But in more complicated engines it acts merely as a signaling device to some other agent that actually manipulates the valve. We can imagine it being replaced by a wire attached at one end to a device which senses when gas is to be expelled from a cylinder and, at the other end, to a motor which suitably actuates the valve. Many modern automobile engines are equipped with electronic fuel injection systems that work very much like this.

There is, however, a limit to the number of mechanical linkages in an automobile engine that can be replaced by information-transmitting devices. The engine ought, after all, to deliver machanical power to the wheels of the car. This requirement places the engine's designer under severe constraints. An engineer may very well conceive an internally consistent set of laws, in other words, a design, for an engine that can nevertheless not be realized. His design might require the machining of metals to tolerances that are simply not achievable with the techniques available to him, for example. Or the strengths of the materials required by his engines may not be realizable with the then-available technologies. But much more importantly, his design may be unrealizable in principle because it violates physical law. This is the rock on which, for example, all perpetual-motion machines will always crash. The laws embodied by a machine that interacts with the real world must perforce to be a subset of the laws governing the real world.

It is, of course, nonsensical to speak of an embodied machine, one made of material substance, that does not interact with the real world. Were such a thing to exist, we could have no knowledge of it—for, in order for it to give evidence of its existence to us, it would have to affect our senses, hence to interact with the real world. In any case, such a machine would be of no use, for by "use" do we not mean interaction with the world?

But there are circumstances under which it is sensible to speak of aspects of real machines that are separate from the machines' physical embodiments. We sometimes need, for example, to discuss what a machine, or a part of a machine, is to do, quite apart from any consideration of how, or of what materials, one might build a device to actually perform the desired action. For example, some part of a gasoline engine must sense when a cylinder's exhaust valve is to be opened and when closed. That function may be realized by a rigid tappet rod or, as I have said, by a wire suitably connected to a sensor and a motor. The rule that such a device is to follow, the law of which it is to be an embodiment, is an abstract idea. It is independent of matter, of material embodiment, in short, of everything except thought and reason. From such a rule, or "functional specification," as engineers like to say, any number of designs may be evolved, e.g., one may be of a mechanical and another of an electrical "tappet rod." The design of a machine is again an abstraction. A good design, say, of a sewing machine, could be given to several manufacturers, each of whom would produce sewing machines essentially indistinguishable from one another. In a sense then such a good design is an abstract sewing machine. It is the sewing machine which could be manufactured—minus, so to say, the material components, the hardware, of the actual sewing machine. The design is also independent of the medium in which it may be recorded. The blueprint of a machine is not its design. If it were, then the design would change whenever the blueprint is enlarged or redrawn in another color. No, a design is an abstract idea, just as is a functional specification. And ideas, say, the idea of a perpetual-motion machine, are not bound by the laws of physics.

Science-fiction writers are forever coming up with, in effect, functional specifications for machines that may be physically unrealizable in that they would violate insurmountable physical principles. One idea that crops up over and over again is that of instant communication over vast distances. Physics has it, however, that no messages in any form whatever can be sent from place to place at a speed greater than that of light. Since the speed of light is finite (approximately 186,000 miles per second), instant communication over even short distances is impossible—at least according to modern physics. Are such ideas as are given us by science-fiction writers therefore useless? No. For although our bodies must function in a world constrained by natural law, our minds are free to leave it. We can give play to our ideas in a world constructed as if the finiteness of the speed of light, for example, were no barrier to the speed of communication generally. To assert that is to say no more than that we may play games whose rules we make up ourselves. We may determine the extent, if any, to which the rules of our games are to correspond to any laws we may think govern the real world. The game "Monopoly" could exist even in worlds, if such there be, in which greed is not a fact.

A crucial property that the set of rules of any game must have is that they be complete and consistent. They must be complete in the sense that, given any proposal for action within the game, they are sufficient for deciding whether that action is legal or not. They must be consistent in the sense that no subset of the rules will determine that a particular action is legal while at the same time another subset determines that that same action is not legal.* A purely abstract game is one whose rules imply no contact with the real world, i.e., one that can be played out in the mind alone. Tournament chess is not such a game, because its rules limit the amount of time a player can devote to considering his moves. This mention of time puts chess in contact with the real world and thus spoils the purity of its abstractness. Apart form that condition, however, chess is a purely abstract game. Another way to state the condition that the rules of a game must be complete and consistent in the sense here intended, is to say that no two referees faced with the same game situation would fail to agree in their judgment. Indeed, "judgment" is not the proper word, for their decision would be reached by the application of logic only. It would, in effect, be nothing more than a determined calculation, a logical process which could have only one outcome.

There is only one kind of question that could reasonably be given for adjudication to a referee of a purely abstract game. A player could describe the game situation, say, the configuration of pieces on a chess board before the disputed action was taken, and

---

* There are, of course, many games whose rules have never been proved to be either consistent or complete in the sense here intended. When, in the playing of such games, difficulties arise because of conflicting or incomplete rules, they are usually resolved by amending the then-known rules. After a time, the so-amended set of rules are thought of as being "classical." (I owe this observation to Oliver Selfridge.)

again after that action. His question must be whether or not it is possible to get from the earlier configuration to the later one in one "move." A player might say, for example, "I had black in check and he castled. Is he allowed to do that?" Or, "I played a spade and he trumped me with a heart, even though he had spades in his hand. Can he do that?" For the rules of an abstract game say only what game configurations can be reached from what other game configurations in a single play or move, and, in some cases, what constitutes a winning configuration. We can say this more technically if we speak of a game configuration as a *state of the game* or, even more simply, as a *state*, and of reaching a state from another state as a *state transition*. Using this terminology, we may characterize the rules of an abstract game as *state-transition rules*.

All games that are interesting to play have permissive state-transition rules. The rules permit the player to make one of a sometimes large number of moves when it is his turn to play, except, of course, in relatively rare forced-move situations. Were that not so, the game as such would be pointless; its whole course, hence its outcome, would be determined even before play began. Still, it may be that the outcome, although determined, is not known to the player, and that he desires to know it.

One may wish to know what time it will be, say 22 hours after 9 o'clock in the morning. To find that out, one will have to play a simple version of the mathematical game called "modular arithmetic." One way to state the fundamental rule of that game is to say that "x mod z" means the remainder when x is divided by z. In the specific example at hand, our player would want to know what the clock would read 22 hours after 9 o'clock, i.e., at 31 o'clock. His problem would be to compute 31 mod 12. (The answer is 7 o'clock.) But let us really make a game of this. The board consists of a set of 12 initially empty ashtrays arranged sequentially (see Figure 2.1). A large number of pebbles is supplied. The player begins by placing as many pebbles as "what time it is now"—9, in our example—in a pile. He then adds to that pile the number of pebbles corresponding to "the number of hours from now" he has in mind—22, in our example. He then selects one pebble from the pile he has just made and places it in the first ashtray. Then, taking another pebble from

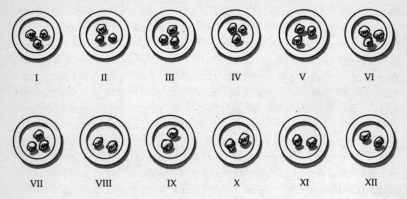

Figure 2.1. The ashtray game for telling time.

the pile, he places it in the next ashtray, and so on until he has either exhausted the pile of pebbles or placed a pebble in the last ashtray. If, when he reaches the latter state, his pile is not exhausted, he repeats the procedure just described. He will eventually have exhausted the pile he made initially. At this point, the last rule is invoked, namely: if any of the ashtrays are empty, his answer is the number of ashtrays that are not empty; if all ashtrays are empty, his answer is 12 o'clock; but if all ashtrays have at least one pebble in them, he take one pebble from each ashtray, and then proceeds to apply the last rule again.

Of course, all this is merely a longwinded game for adding two numbers and then dividing their sum by 12 by means of successive subtraction. The rules of the game are not permissive; they don't allow the player to choose the transition from one state of play to the next from a number of alternatives. To the contrary, they command precisely what he must do to make that transition. Such a set of rules—that is, a set of rules which tells a player precisely how to behave from one moment to the next—is called an *effective procedure*. The notion of an effective procedure, or "algorithm," as it is also called, is one of the most important in modern mathematics. Not only is much of mathematics concerned with finding effective procedures for doing all sorts of useful things, long division, for example, but there exist deep mathematical questions, having to do with the fundamental nature of mathematics itself, that become statable and attackable when formulated as questions about effective procedures.

The definition of an effective procedure given above is deceptively simple. The deception is in the words "tell a player." A player who undertakes to solve his time-telling problem by following the rules I have just stated must first understand those rules. He must know what it is to make a pile of pebbles, what an ashtray is, how to tell when an ashtray is empty (suppose it contains ashes, but no pebbles), and so on. He must, in other words, be able not only to read the rules, but to interpret them in precisely the way I intended them to be interpreted. And if the rules are to tell a player "precisely" how to behave, then the rules must be expressed in a language capable of making precise statements. Are cookbook recipes, for example, effective procedures? They certainly attempt to tell a cook what to do from one moment to the next. But then they are generally, even usually, laced with phrases such as "add a pinch of paprika," "stir until consistent," and "season to taste." Would we not all agree that such directions are far from precise? Yet we can imagine a cooking academy that trains its students to such a high standard of both performance and taste discrimination that even such directions, vague as they are to the ordinary person, have a precise and uniform meaning for them. That school's recipes would then seem to constitute effective procedures for its graduates, though not necessarily for anyone else.

My characterization of an effective procedure as a set of rules which tells a player precisely how to behave from one moment to the next appears, then, to be defective, at least in that it can't stand on its own legs. Given a set of rules, say for baking a cake, we appear to have no absolute criteria for determining whether it is or is not an effective procedure. There would be no such difficulty, at least not for cooking recipes, if two conditions were fulfilled: first, that there exists a language in which precise and unambiguous cooking rules could be stated; and, second, that all people are identical in every respect having anything to do with cooking. These conditions are not independent of one another, for one way in which everyone would have to be like everyone else is that they would all have to interpret the cooking language identically. But even the most rigorous cooking academies do not demand that their students become exactly like their master in all relevant respects. They hope only that their graduates have learned to *imitate* the master chef in his interpretation of recipes.

This display of modesty on the part of the senior faculties of cooking schools serves us as an example from which we may learn how to proceed along our own way. In order to give the notion "effective procedure" autonomous status, we need a language in which we can express, without any ambiguity whatever, what a player is to do from one moment to the next. But allegedly effective procedures may be written in languages, many of which are, unless constrained by specially constructed rules, inherently ambiguous.

The problem that thus arises would be solved if there were a single inherently unambiguous language in which we could and would write all effective procedures. It would be sufficient if we used that language, not for writing effective procedures we wish to execute, but for writing rules for interpreting other languages in which such procedures may actually be written. For if an agent competent in only one language were given a procedure written in a language strange to him, together with rules that dictate precisely how to interpret statements in the strange language, then he could imitate what the behavior of a speaker of the strange language would have been had that speaker followed the given procedure. We need therefore some absolutely unambiguous language in which we can write effective procedures and in terms of which we can state rules for interpreting statements in other languages. Such sets of rules would again have to be effective procedures, namely, procedures for the interpretation of sentences of the language to which they apply. But in what language are these rules to be written? We appear to have entered an infinite regress. Had we such a language, and we shall see that we can construct one, we could say of every procedure written in an unambiguous language precisely what it tells us to do: do what the imitating agent does. Hence every such procedure would have a unique interpretation that is independent of the language in which the procedure was originally written.

I have, by virtue of my silence on the point, let stand the impression that whenever I refer to languages I mean not only formal languages like that of arithmetic, but also natural languages like English and German. Indeed, I stated the rules of the time-telling game in English but also mentioned arithmetic. Yet we demand of the languages we have just discussed that they have unambiguous rules of interpretation. We know that natural languages are notorious for their ambiguity. Later on we will consider what it means to "understand" natural languages in formal terms. But for the moment let us restrict our attention to formal languages.

A formal language is (again!) a game. Let us return briefly to a consideration of the game of chess. It consists of a set of pieces, a board having a certain configuration, a specification of the initial positions of the pieces on the board, and a set of transition rules which tells a player how he may advance from one state of the game to the next. We have already noted that these rules are, except under certain circumstances, permissive; they tell the player the moves he *may* make, but don't dictate what he *must* do. The fact that the initial state of the chess game is specified is a peculiarity of chess, not a reflection of a property of games generally. In poker, players are dealt five cards each, but there is no specification for what these cards must be. Of course, every game must be initialized somehow. We may as well speak of the initialized game as being its starting state, and then add, to the already existing state-transition rules of the game, formation (as opposed to transformation) rules, sometimes permissive as in poker and sometimes mandatory as in chess; formation rules tell how to transform that beginning state into what we ordinarily think of as the initial state of the game, e.g., the cards dealt out or the pieces set up on the board.

The "pieces" of a formal language are its *alphabet,* i.e., the set of symbols which may be manipulated in the language. We may, if we wish to preserve the analogy to chess, think of the paper on which the symbols of the language are written as the "board," but that is not important. The transition rules of a formal language play the same role for it as the transition rules of a game play for the game: they tell a player how to move from one state of the game to another. I said earlier that the only significant question that can be put to a game's referee is whether a proposed move is legal or not. Exactly the same is true for formal languages. However, with formal languages, although strictly speaking only that one question is possible, it can take two different forms: first, "Is the proposed transition legal?" and second, "Is the configuration of symbols under consideration an admissable expression in the language?" "May I castle while my king is in check?" is a chess question of the first form. "Is the board configuration here exhibited one that could possibly be reached by legal play?" would be a chess question of the second form. For some board configurations that question is easy to answer. Were we asked it about a board, for example, on which there were eight white pawns and on which two white bishops occupied squares of the same color, we would answer "No." Similarly if we found two kings of the same color on the board, and so on. But such questions are simply not asked by chess players. The reason for this is that a chess game always starts from the standard board configuration or is resumed from a position achieved by a temporarily suspended game.

Many formal languages differ from chess in this respect. High-school algebra—whose rules I will not detail here—has, for example, transformation rules for factoring algebraic expressions; e.g., $ac + bc$ is transformed into $(a + b)c$ by one such rule. But in order for such rules to be applicable at all, the expressions to which they are to be applied must first of all be legal (grammatical) expressions (or sentences) in the language. The expression $ac + bc +$ is not a correct sentence in algebra and none of algebra's transformation rules apply to it. If one is to play algebra, then, one must first set up the board in a legal manner. One must know that an expression beginning with a left parenthesis must somewhere be "closed" by a matching right parenthesis, that operator symbols like "$+$" must be placed between two expressions, and so on.

A formal language is a game. That is not a mere metaphor but a statement asserting a formal correspondence. But if that statement is true, we should, when talking about a language, be able to easily move back and forth between a game-like vocabulary and a corresponding language-like vocabulary. Precisely that can be done.

I will describe a game in terms of a board, pieces, moves, and so on, and finally develop the corresponding linguistic notions. My purpose is to develop a very precise language on a very small alphabet, moreover, a language whose transformation rules can also be written using only that small alphabet. I will sketch the design of a simple machine embodying those transformation rules. That machine will, of course, be able to play the game I described initially. I will then write the rules of the game I have designed in a language whose alphabet corresponds to the game's pieces. It is that language that is the game. Finally, I will, indicate how a second machine can be designed to function as an interpreter of this new language. We shall see where we go from there.

THE GAME

Equipment:

> One roll of toilet paper.
> Many white stones, five black stones, and an
> > ordinary six-sided playing die.

Initialization:

1. Roll out the toilet paper on the floor.
2. Put down stones as follows:
   i. on an arbitrary square, one black stone;
   ii. on successive squares to the right of the square holding the black stone; as many white stones as you please, one to each square;
   iii. on successive squares continuing to the right, one black stone, skip one square, one black stone;
   iv. on successive squares continuing to the right, an arbitrary number of white stones;
   v. on the square to the right of the last white stone, one black stone;
   vi. finally, one black stone, the "marker," above the square holding the rightmost white stone.
3. Turn the die so that its one-dot side is facing upward, i.e., so that it is showing "1."

(An example of such a setup is shown in Figure 2.2.)

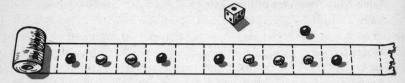

Figure 2.2. The initial configuration of the game.

The Transformation Rules:

The marker stone is moved either one square to the left or one to the right on each move. However, before each move, the stone under the marker stone is replaced or removed according to the applicable rule. The die may be turned to a new side after each move.

The rules govern: what kind of stone, if any, is to replace the stone under the marker; what side of the die is to be turned up; and in which direction the marker is to be moved. There are eighteen rules. They all have the same form. Each of them describes an orientation of the die, and a specific kind of stone or no stone at all under the marker. The player must find the rule that corresponds to the existing game situation, i.e., the situation defined by what the die reads and by what kind of stone, possibly no stone at all, is under the marker. He is then to do what the rule tells him to do. Each rule says to do three things:

1. Turn the die so that it reads the stated number.
2. Replace the stone under the marker by the kind of stone specified—possibly by no stone at all.
3. Move the marker one square in the indicated direction.

Obeying a rule thus creates a new game situation. The player again applies the rule then appropriate, and so on. When a rule tells him to turn the die so that it will read "O," the game stops.

The rules are given in the form of a table (Table 2.1). The first two columns describe the conditions under which the corresponding rule is applicable, and the remaining three columns of the corresponding row tell what to do.

Table 2.1. *The Rules of the Game.*

IF THE DIE READS	AND THE STONE UNDER THE MARKER IS	THEN TURN THE DIE TO	REPLACE THE STONE BY	MOVE MARKER
1	none	3	white	left
1	black	2	no stone	left
1	white	1	white	left
2	none	2	no stone	left
2	black	3	no stone	left
2	white	5	no stone	right
3	none	3	no stone	left
3	black	4	no stone	right
3	white	5	no stone	right
4	none	4	no stone	right
4	black	1	black	right
4	white	6	white	left
5	none	5	no stone	right
5	black	1	black	right
5	white	1	white	left
6	none	0	no stone	right
6	black	0	black	right
6	white	3	white	left

Of course, just as Norwegian sardines in the Second World War were not for eating but for buying and selling, so this game is not for playing but for talking about. In order to be able to talk about it more easily, let us change notation: instead of "black," "white," and "no" stones, we will "X," "1," and "0," respectively. An initial board configuration is then

$$\ldots 000X11X0X111X00\ldots,$$

where "..." means "and so on." (Remember, the whole roll of toilet paper contains "0," i.e., no stones, initially.) The marker is really only an aid to memory; it can be replaced by the player's index finger. I have here underlined the marked place. If we now interpret rows of 1's as numbers—"111" means "3"—we can see that the X's serve as punctuation, rather like quotation marks, enclosing two numbers. Given the above initial configuration, the play would end with the configuration

$$...00011111X00....$$

This configuration may be interpreted as the sum of the two numbers initially presented. The game thus constitutes (although I haven't here proved it) an adding machine. The whole game is shown in Table 2.2, in which the rightmost column gives the number displayed by the die.

As an example of an application of a rule, consider the transition from line 6 to line 7 in the game shown. Line 6 is

$$X11\underline{X}00111X$$

with the die reading 2. The applicable rule must therefore say "If the die reads 2 and a stone of kind X, i.e., a black stone, is under the marker, then do such and such." The fifth—and only the fifth—of the given rules matches the conditions pertaining to line 6. It says to do as follows:

1. Turn the die so that it reads 3.
2. Replace the black stone under the marker by no stone; i.e., remove the black stone.
3. Move the marker one square to the left.

After that rule has been followed, the board configuration of the game is then the one displayed in line 7,

$$X11\underline{0}00111X,$$

and the die is left reading 3.

Now look at the play of this game as a whole, and notice particularly how the marker shuffles back and forth. To aid the intuition, think of the game being played on a field. The stones are very heavy and the boy moving them must rest each time he moves from

Table 2.2. *The Game.*

STEP	BOARD	NEXT RULE
1	X11X0X111X	1
2	X11X0X111X	1
3	X11X0X111X	1
4	X11X0X111X	1
5	X11X00111X	2
6	X11X00111X	2
7	X11000111X	3
8	X10000111X	4
9	X10000111X	5
10	X10000111X	5
11	X10000111X	5
12	X10000111X	1
13	X10001111X	3
14	X10001111X	3
15	X10001111X	3
16	X10001111X	3
17	X00001111X	5
18	X00001111X	5
19	X00001111X	5
20	X00001111X	5
21	X00001111X	1
22	X00011111X	3
23	X00011111X	3
24	X00011111X	3
25	X00011111X	3
26	000011111X	4
27	000011111X	4
28	000011111X	4
29	000011111X	4
30	000011111X	6

one position to the next. We can see the major outlines of his strategy: He searches for the rightmost "1" of the left number (and finds it in step 7) and then for a place to put it. To find that, he must find the leftmost "1" of the right number (which he finds in step 11) and replace its left neighbor by "1." He continues in this fashion until he runs into the leftmost "X," a boundary marker, and concludes he is done. Of course, he may not see this strategy. He carries the rules with him and consults them between each move.

It is easy to see how we could build a machine to do this work. We replace the roll of toilet paper with an ordinary reel of magnetic tape, and the player with an ordinary tape recorder. Of course, we have to change the rules a little: now the tape moves, not a marker along the tape. Therefore whenever we specified a marker motion to the left, we must now specify a tape motion to the right, and vice versa. The three symbols we need can be represented by three easily distinguishable tones. We install six relays in the tape recorder to hold the information formerly conveyed by the die. (Actually, three relays would do, but imaginary relays are cheaper than complicated explanations.) Let a very high tone stand for "X," a medium tone for "1," and a very low tone for "0." If we have recorded a tape to correspond to the initial configuration given in Table 2.2, we can start the computation by placing the tape in the machine so that the rightmost "1," i.e., medium tone, is under the read/write heads of the recorder and by seeing to it that relay 1 is closed—because the die showed 1 initially—and that all other rule relays are open. The circuitry of the machine is so arranged that:

> If relay 1 is closed and if a high tone (X) is read, then the high tone is removed by overrecording a low tone ("0"), the tape is moved one square to the right (this effectively moves the recorder's heads, i.e., the marker, to the left), relay 1 is turned off (opened), and relay 2 is turned on (closed).

This is a faithful translation of the second line of our original table of rules. Similarly, every other one of our rules is translated into tape-recorder terms. We then have an incredibly awkward adding machine.

The most important thing to notice about this machine is that it is completely defined by the rules of the game as translated into tape-recorder terms. Given those rules, it is an adding machine; given some others, it would be some other kind of machine.

Now notice also that we needed only three distinguishable symbols on our tape. We denoted them by "X," "1," and "0." It turns out that we can write our entire set of rules using only these three symbols as well!

Let us speak of the "state" of the machine as being the number showing on its die, i.e., the number of the rule relay that happens to be on (closed). We now write the five parts of our rules in the following sequence,

Current state, symbol under heads, next state, symbol to be written, direction of tape motion,

and we adopt the following code:

1. "X" is a punctuation mark indicating the beginning and end of an item of information. It is, in other words, a kind of bracket.

2. When we wish to indicate a number, we write the corresponding number of 1's.

3. In the context "Direction of tape motion," "0" stands for "left" and "1" for "right."

Using this code we may now transliterate the rules given in Table 2.1 into the form shown in Table 2.3.

We have written these rules on separate lines, and we have separated columns by blank spaces. No confusion could result if we eliminated the blanks and concatenated the lines into one long string of X's, 1's, and 0's. That string would then constitute a complete description of our adding machine!

We have now developed a notation in which we can describe a machine. Its alphabet consists of the three symbols "X," "0," and "1." Of course, strings written in this notation would remain meaningless unless we could also say how they are to be interpreted. To

Table 2.3. *The Rules of the Game.*

RULE NUMBER	SYMBOL UNDER HEAD	NEXT RULE NUMBER	SYMBOL TO WRITE	DIRECTION OF TAPE MOTION
X1X	X0X	X111X	X1X	X1X
X1X	XXX	X11X	X0X	X1X
X1X	X1X	X1X	X1X	X1X
X11X	X0X	X11X	X0X	X1X
X11X	XXX	X111X	X0X	X1X
X11X	X1X	X11111X	X0X	X0X
X111X	X0X	X111X	X0X	X1X
X111X	XXX	X1111X	X0X	X0X
X111X	X1X	X11111X	X0X	X0X
X1111X	X0X	X1111X	X0X	X0X
X1111X	XXX	X1X	XXX	X0X
X1111X	X1X	X111111X	X1X	X1X
X11111X	X0X	X11111X	X0X	X0X
X11111X	XXX	X1X	XXX	X0X
X11111X	X1X	X1X	X1X	X1X
X111111X	X0X	XX	X0X	X0X
X111111X	XXX	XX	XXX	X0X
X111111X	X1X	X111X	X1X	X1X

do this, it would be sufficient to describe a machine—or, better yet, to build one—that would take as its input the two streams of information consisting of, first, the properly encoded description of our adding machine, and, second, an initial configuration of X's, 0's, and 1's on which the so-described adding machine is to operate. These two so-called *strings* of symbols could, of course, be put on a single tape. Once we have such a machine, we are entitled to call our system of notation a language, for we will then have an embodiment of its transformation rules.

In one of the greatest triumphs of the human intellect, the English mathematician Alan M. Turing proved in 1936 that such a machine could be built and even showed how to build it.[1] He actually proved much more—but of that, more later. I cannot here describe a machine built according to Turing's principles in any great detail, but I must say a few words about it.

Imagine again a tape recorder such as the one we used for our adding machine. Again it has a set of relays capable of representing its states. This time, because the machine is much more complicated than our adding machine, there are many more states to be represented and hence more relays—but that is a detail. The tape we give our new machine has the following layout; reading from right to left:

> a section containing the description of our adding machine in the notation we have developed;
>
> a section containing the data on which the adding machine is to work, for example, "X11X0X111X";
>
> a section that can store the current state of the "adding machine";
>
> an arbitrarily long section of "blank" tape, i.e., a section containing 0's.

The information on the tape then has the following structure:

blank tape	current state	current symbol	data	machine description

To see how the machine works, one need only pretend that one has been given a description of the adding machine, the relevant data set, and a few conventions, such as that the adding machine always starts off in state 1 and that the "X" at the extreme right of its data set is a marker indicating the beginning of data to the left. That plus some scratch paper. This is precisely what the machine has. It shuttles the tape back and forth, reading data, making marks on its scratch tape, and inspecting the machine-description portion of its tape for information on what to do next. It thus slowly, very slowly, but step by step and utterly faithfully, imitates our original adding machine, i.e., the machine described to it on its tape. This machine then truly embodies the rules telling how to interpret the strings of X's, 1's, and 0's we have given it.

A machine of the kind we have been discussing, i.e., a machine that shuttles a tape back and forth, reading and changing marks on a square of tape at a time and going from one state into another, and so forth, is today called a Turing machine. Such a machine is completely described when, for every state it can attain and for every symbol that can be under its reading head, the description states what symbol it is to write, what state it is to go into, and in which direction it is to move its tape. We can stipulate, as a matter of convention, that every such machine always start in its first state, call it state 1, and with the tape so positioned that its rightmost symbol is under the read/write head. We happened here to use a language whose alphabet consists of the three symbols we used to describe our adding machine. Had we been more generous, i.e., had we permitted ourselves the use of a larger alphabet, our machine could have been simpler, in the sense of having fewer states. On the other hand, the rules would have been more complicated. There are, then, many possible realizations of our adding machine—at least one for every substantially different alphabet we could have chosen. The minimum size of an alphabet we could use is two—an adding machine such as ours would have to have many states to operate with such a restricted alphabet. The minimum number of states such an adding machine would have to have is two—but this machine would have to operate on a large alphabet.

There is an important difference between the two machines we have been discussing: our adding machine is a special-purpose machine. It can add any two numbers, but it can do nothing else. The second machine requires as input an encoded description of some machine and a data set on which the described machine is to operate. In effect, the machine description it is given is a *program* which transforms the second machine into the machine it is to imitate. The question naturally arises "What kinds of machine can be imitated in this way?"

I have illustrated what I mean by a formal language, namely: an alphabet; a set of formation rules that determine the format of strings of symbols constructed on that alphabet, that constitute legal expressions in the language; and a set of transformation rules for such expressions. I have also said that it is possible to construct machines that embody such transformation rules and that can therefore execute procedures represented in the corresponding languages. Beyond that, I have discussed a machine that accepts descriptions of other machines and is capable of imitating the behaviors of the described machines. We have thus gained an idea of what is meant by imitation in the present context. Note carefully that I have never alluded to translation of one language to another. Our imitating machine does not first translate the transformation rules we gave it, i.e., the encoded description of our adding machine, into its own or any other language. It consults—we use the word "interpret"—that set of transformation rules each time it must decide what the machine it is imitating would do. It thus makes many moves for every move the imitated machine would have made.

My aim, for the moment, is to put a firm foundation under our concept of effective procedure. We wish for a single language in terms of which effective procedures can be expressed, at least in the sense that we can describe all our procedural languages in that language and thus give our procedures unique interpretations. We can now see that the transformation rules of languages can be embodied in machines. My task is therefore reduced to showing that a unique alphabet, and a language on that alphabet, can be found in which we can indeed describe all languages in which we may want to write procedures.

We can design a language whose alphabet consists of only two symbols, say, "0" and "1," in terms of which we can describe any Turing machine. Now we have seen that a language consists of more than an alphabet, just as a game consists of more than the pieces with which it is played. Its transformation rules must also be given. In the present context I intend the transformation rules of the language I have in mind to be embodied in a Turing machine similar to the imitating machine we have already discussed. I am saying, then, that there exists a Turing machine that operates on a tape containing only 1's and 0's and that is capable of imitating any other Turing machine whatever. This so-called *universal* Turing machine is, as are all Turing machines, describable in terms of a set of quin-tuples of the form with which we are already familiar, i.e.,

(current state, symbol read, next state, symbol
written, direction of tape motion),

and these quintuples, in turn, may be written in the language that that Turing machine is designed to accept. The universal Turing machine is consequently capable of accepting a description of itself and of imitating itself.

In fact, one can design many languages with the same two-symbol alphabet, i.e., many universal Turing machines embodying rules of transformation on strings of these two symbols. And one can, of course, enlarge the alphabet and design many universal Tur-ing machines corresponding to each such enlargement. But it is the principle that interests us for the moment, namely, that

there exists a Turing machine U (actually a whole class of ma-chines) whose alphabet consists of the two symbols "0" and "1" such that, given any procedure written in any precise and unam-biguous language and a Turing machine L embodying the transfor-mation rules of that language, the Turing machine U can imitate the Turing machine L in $L's$ execution of that procedure.

This is a restatement of one of the truly remarkable results that Turing announced in his brilliant 1936 paper.

There are many existence proofs in mathematics. But there is a vast difference between being able to prove that something ex-ists and being able to construct it. Turing proved that a universal Turing machine exists by showing how to construct one. We have to remember that Turing did this monumentally significant work in 1936—about a decade before the first modern computers were actu-ally built. Modern computers hardly resemble the machine Turing described. Many have, for example, the ability to manipulate many magnetic tapes simultaneously and, even more importantly, most are equipped with very large information stores. The storage mecha-nism of a modern computer is functionally like a set of relays, each of which can be either on (closed) or off (open). A set of ten such relays can take on 1,024 different states. It is not uncommon for a modern computer of moderate size to have more than a million such elementary storage components, and thus to be able to take on $2^{1,000,000}$ states. That is an unimaginably huge number. (The Earth, for example, weighs much less than $2^{1,000}$ pounds.) Still, in principle, every modern computer is a Turing machine. Moreover, every modern computer, except for very few special-purpose machines, is a *universal* Turing machine. And that, in practice, means that every modern computer can, at least in principle, imitate every other modern computer.

There is still one more hole to be plugged. For even granting that, in effect, any computer can do what any other computer can do, there remains the question of what computers can do at all, i.e., for what procedures one can realize Turing machines, hence Turing machines imitable by universal Turing machines, and hence imitable by modern computers. Turing answered that question as well: a Turing machine can be built to realize any process that could naturally be called an effective procedure.

This thesis, often called Church's thesis after the logician Alonzo Church, who formulated it in a framework different from Turing's, cannot be proven, because it involves the word "naturally." In a sense, we are stuck in a logical circle; any process we can describe in terms of a Turing machine is an effective procedure, and vice versa. What lends real intuitive strength to the idea, however, is the fact that several radically different and independently derived formulations of the idea of "effective computability" have all been shown to be equivalent to computability in Turing's formalism and hence to one another. As M. Minsky remarks; "Proof of the equivalence of two or more definitions always has a compelling effect when the definitions arise from different experiences and motivations."[2]

But even though we must rely on our intuition as to what may "naturally" be called an effective procedure, we are now on firm ground in being able to say precisely and unambiguously what an effective procedure tells us to do. At least in principle, we can encode the alphabet of the language in which the procedure is written, using only the two symbols "0" and "1." We can then transcribe the rules that constitute the procedure in the new notation. Finally, we can given some universal Turing machine that operates on the 0,1 alphabet the transformation rules of the procedure's language in the form of suitably encoded quintuples. The given procedure tells us to do what the so-instructed universal Turing machine does as it imitates the "machine" we have described to it. If we understand how a Turing machine operates at all (and such understanding involves very little knowledge, as we have seen), and if we have a description of the universal Turing machine we appealed to, then we know what the procedure tells us to do in detail.

Such a way of knowing is very weak. We do not say we know a city, let alone that we understand it, solely on the basis of having a detailed map of it. Apart from that, if we understand the language in which a procedure is written well enough to be able to explicate its transformation rules, we probably understand what rules stated in that language tell us to do.

But such objections, valid as they are, miss the point. Turing's thesis tells us that we can realize, as a computer program, any procedure that could "naturally" be called an effective procedure. Therefore, whenever we believe we understand a phenomenon in terms of knowing its behavioral rules, we ought to be able to express our understanding in the form of a computer program. Turing proved that all computers (save a few special-purpose types that do not concern us here) are equivalent to one another, i.e., are all universal. Hence any failure of a technically well-functioning computer to behave precisely as we believe we have programmed it to behave cannot be attributed to any peculiarity of the specific computer we have used. Indeed, the fault must be that we have been careless in our transcription of the behavioral rules we think we understand into the formal language demanded by our computer, or must be in the initial explication, in any form, of what we had in mind when we believed we understood, or must be that our understanding is defective. The last is most often the case. I shall say much more about that later. For now, we need note only that the defect in our understanding can take two forms:

First, although our theory may be on the whole correct, it may contain an error in detail. We wrongly assert, for example, that if this and that is true, then so-and-so follows. Our mental processes, lulled perhaps by the sheer eloquence of the argument we make to ourselves, often permit us to slide over such errors without the slightest disturbance. The computer is, however, very unforgiving in this respect. It follows the logic we have given it. That logic may lead to very different consequences than do mental processes contaminated by wishes to reach certain outcomes. Indeed, one of the most cogent reasons for using computers is to expose holes in our thinking. Computers are merciless critics in this respect.

Second, the defect in our understanding may be that, although it is true that we understand, we are still not able to formalize our understanding. We may, for example, be able to predict with very great confidence what an animal will do under a large variety of circumstances. But our predictive power, great and reliable as it may be, may rest on intuitions that we are simply unable to adequately explicate. Yet we may be driven to force our ideas into a formal mold anyway. A computer program based on a formal system so derived is certain to misbehave. The trouble then is not merely that the theory it represents contains certain errors in detail, but that that theory is grossly wrong in what it asserts about the matters it concerns. It is not always clear which defect one is confronted with when a computer one has programmed misbehaves. There is usually enormous motivation to believe that one's theories are all right on the whole, and that, when they don't work well, there must be some error in detail that can easily be patched up. I shall have more to say about such matters later.

We have used the idea of the universality of computers in the foregoing. We must now ask whether the universality of computers implies that they can "do anything." This is really the question "Can anything we may wish to do be described in terms of an effective procedure?" The answer to that question is "No."

First, there are certain questions that can be asked and for which it can be proved that no answers can be produced by any effective procedure whatever. We may, for example, be interested to know whether some machine we have designed, say, our adding machine, will halt once started with a particular data set. It would be convenient if we had a testing machine which could, for any machine and any data set appropriate to it, tell us whether that machine operating on the given data set would ever halt. No such machine can be built. This and many other such "undecidable" questions therefore impose some limit on what computers can do. Of course, this is a logical limitation, which constrains not only electronic computers but every computing agent, human and mechanical. It has also to be said that the whole set of undecidable questions is not terribly interesting from a practical point of view; all such questions are vastly general. If we had some specific computation about which we wanted to know whether or not it would ever terminate, we usually could design a procedure to discover that. What is impossible is to have a machine—or, what is the same thing, an effective procedure—that will make that discovery for any procedure *in general.*

Second, an effective procedure may be capable of making some calculation in principle, but may take such a long time to complete it that the procedure is worthless in practice. Consider the game of chess, for example. Given the rule that a game is terminated if the same board configuration is achieved three times, chess is certainly a finite game. It is therefore possible, in principle, to write a procedure to generate a list of all games, move for move, that could possibly be played. But that computation would take eons to complete on the fastest computers imaginable. It is therefore an example of an impractical procedure. Indeed, we have discussed procedures up to this point as if the time they take to do their work, i.e., to complete their computational task, were irrelevant. Such an attitude is appropriate as long as we are in the context of abstract games. In practice, of course, time does make a difference. We must note in particular that, when one computer is imitating another, it must go through many time-consuming steps for every single step of the imitated computer. Were that not so, we would strain to build the cheapest possible universal Turing machine and, since it could imitate every more expensive machine, it would soon drive all others from the market.

Third, we may write a procedure realizable by a Turing machine, hence an effective procedure, but one whose rules do not include an effective halting rule. The procedure, "beginning with zero, add one, and, if the sum is greater than zero, add one again, and so on," obviously never stops. We could substitute "if the sum is less than zero, stop, otherwise add one again" for "if the sum is greater than zero, add one again" in that procedure and thus provide it with a halting rule. However, a computation following that procedure would never encounter the halting rule, i.e., the corresponding Turing machine would never fall into the state corresponding to "sum less than zero." The procedure is therefore, in a sense, defective. It is not always easy, to say the very least, to tell whether or not a real procedure written for real computers is free of defects of this and similar kinds.

Finally we come to the most troublesome point concerning what computers can and cannot do. I have said over and over again that an effective procedure is a set of rules which tells us in precise and unambiguous language what to do from one moment to the next. I have argued that a language is precise and unambiguous only if its alphabet and its transformation rules can themselves be explicated in precise and unambiguous terms. And I have repeated Church's (and Turing's) thesis that, to every such explication of whatever language, there corresponds a Turing machine that can be imitated by a universal Turing machine. I have asserted further that virtually every modern computer is a universal Turing machine. Leaving to one side everything having to do with formally undecidable questions, interminable procedures, and defective procedures, the unavoidable question confronts us: "Are all the decisionmaking processes that humans employ reducible to effective procedures and hence amenable to machine computation?"

We have seen that the very idea of an effective procedure is inextricably tied up with the idea of language. Isn't it odd that I could have spent so much time discussing language without ever alluding to *meaning*? The reason I have been able to avoid confronting the concept of meaning is that I have been discussing only formal languages or, as I have said, abstract games. Not that meaning plays no role whatever in such language games. It does. But this role is entirely subsumed in the transformation rules of the language. Recall that in algebra we may transform $ac + bc$ into $(a + b)c$. We are entitled to say that the two expressions mean the same thing, or, to put it another way, that the transformation we have employed preserves the "value" of the original expression. In still other terms, were we to substitute numbers for *a, b,* and *c,* the two expressions would both produce the same result upon execution of the indicated arithmetic operations. (This last is, by the way, not a property of all algebras. Elementary algebra has been deliberately designed so that its transformation rules are consistent with those of the formal language we call arithmetic.) It is a property of formal languages, indeed, it is their essence, that all their transformation rules are purely syntactic, i.e., describe permissable rearrangements of strings of symbols in the language, including replacements of symbols and introductions of new symbols—e.g., ")" and "("—independent of any interpretation such symbols may have outside the framework of the language itself. One can, for example, do pages of algebraic transformations, following the rules of algebra blindly, without ever having to know that one may substitute numbers for lowercase letters but not for parentheses, in other words, without ever giving any interpretations to the symbols one is dealing with. The same is not true for natural language. Consider the English sentence: "I never met a man who is taller than John." It may be transformed into "I never met a taller man than John." This transformation clearly preserves the meaning of the original sentence. But if we apply the same transformation rule to "I never met a man who is taller than Maria," and get "I never met a taller man than Maria," it no longer works. The rule we have applied is not purely syntactic. It concerns itself not merely with the *form* of uninterpreted strings of symbols, but with their *meanings* as well.

We have seen that, at a certain level of discourse, there is no essential difference between a language and a machine that embodies its transformation rules. We have also noted that, although the laws of which abstract machines are embodiments need not necessarily be consistent with the laws of the physical universe, the laws embodied by machines that interact with the real world must perforce be a subset of the laws governing the material world. If we wish to continue to identify languages with machines even when discussing natural language, then we must recognize that, whatever machines correspond to natural languages, they are more like machines that transform energy and deliver power than like the abstract machines we have been considering; i.e., their laws must take cognizance of the real world. Indeed, the demands placed on them are, if anything, more stringent than those placed on mere engines. For although the laws of engines are merely subsets of the laws of physics, the laws of a natural-language machine must somehow correspond to the inner realities manifest and latent in the person of each speaker of the language at the time of his speaking. Natural language is difficult in this sense because we have to know, for example, to what "values" of X and Y we can apply the transformation rule that takes us from "I never met an X who is taller than Y" to "I never met a taller X than Y." It is clearly not a rule uniformly applicable to uninterpreted strings of symbols.

This difficulty is even deeper than may be at first apparent. For it is not even possible to define the domain of applicability of this rule—and there are many like it—by, say, using lists of male and female nouns, pronouns, and names to suitably amend the rule. Consider a variation of the very example I have cited, namely, the sentence "I never met a smarter man than George." Imagine a detective story in whose first chapter it becomes clear that the sought-after criminal must be someone who is pretending to be something he is not. The master detective unmasks the imposter in Chapter 10, say. The purpose of Chapter 11 is to explain to the reader who has not been able to infer the detective's conclusions from the clues provided throughout the book, just how the detective came to identify the guilty person. In Chapter 11, then, the detective explains that he overheard Mr. Arbothnot make the remark "I never met a smarter man than George" at a literary tea at which the work of the English author George Eliot was being discussed. Mr. Arbothnot had gained an invitation to that tea by persuading his hostess that he was an authority on nineteenth-century English literature. The detective reasoned that anyone who knew anything about English letters would know that George Eliot was a pseudonym for Mary Ann Evans, a lady. Mr. Arbothnot's remark was therefore "ungrammatical," in somewhat the same way that the mathematical expression x/y is ungrammatical whenever $y=0$; hence Mr. Arbothnot could not be what he claimed to be.

A good detective story, perhaps we should say a "fair" one, is one that gives the reader all the information necessary to discover the truth, e.g., who did it, before explaining how the detective made his deductions. The whole point of a detective story is often just the disambiguation of what one of its characters said in its early parts. The very possibility of spotting an ambiguity, hence of knowing that it requires disambiguation, hence the possibility of solving the mystery, depends on the reader's knowledge of the real world and on the property of natural language that its rules apply to strings of symbols interpreted in real-world contexts. A story of the kind we have discussed cannot be understood in solely formal terms. Interestingly enough, neither can its first chapter be translated into another language without the translator's knowledge and understanding of Mr. Arbothnot's fatal mistake, without, that is, an understanding reading of the denouement provided in the last chapter.

We will return later to considering the role that context plays in understanding natural language—whether by humans or by machine. For now our concern is still with the narrower question—at least narrower as it is here construed—of the convertability of human decisionmaking processes into effective procedures, hence into computable processes.

There are, of course, human decisionmaking processes that can be described clearly and unambiguously even in natural language. I have described games here both in natural language and in terms of machine designs that I again described in natural language. Indeed, we could not understand a Turing machine or an effective procedure cast in Turing machine terms, i.e., as a program for some universal Turing machine, without first understanding what it means for one square on a tape to be adjacent to another, what it means to read and write a symbol or a square of tape, what it means for a tape to be moved one square to the right or left, and so on. What is so remarkable is how incredibly few things we must know in order to have access, in principle, to all of mathematics. In ordinary life we give each other directions, i.e., describe procedures to one another, that, although perhaps technically ambiguous in that they are potentially subject to various interpretations, are, for all practical purposes, effective procedures. They rest at bottom on extremely widely shared vocabularies whose elements, when they appear in highly conventionalized contexts, have effectively unique interpretations. Most professional and technical conversations avail themselves of such vocabularies almost exclusively. The problem of converting such procedures into effective procedures in the technical sense, i.e., into programs for Turing machines, is fundamentally one of formalizing the knowledge base that underlies the conventional interpretation of their vocabularies. The more highly standardized these vocabularies are, and the more restricted the context in which they are used, the more likely that this problem can be solved. For, after all, if each symbol of a set of symbols has an effectively unique interpretation in a certain context, and if strings of such symbols are transformed only by rules that themselves arise out of that context, then no question of giving each symbol an interpretation arises in any formal sense at all. A language so constrained is effectively a formal language. Its rules are therefore potentially realizable by a Turing machine.

But then there remain the many decisions we make in daily life for which we cannot describe any decisionmaking process in clear language. How do I decide what word to write next? Perhaps our incapability in this respect is due entirely to our failure till now to come to an adequate understanding of human language, the mind, the brain, and symbolic logic. After all, since we can all learn to imitate universal Turing machines, we are by definition universal Turing machines ourselves. That is, we are *at least* universal Turing machines. (Even a physically realized Turing machine is not a mere Turing machine; it may, for example, be a bookend or a paperweight as well.)

We join Michael Polanyi in saying that we know more than we can tell.[3] But in so saying we have come full circle. Our question is, "What can one tell computers?" We have taken telling to mean giving an effective procedure. And the question we are presently entertaining is, "Can anything we may wish to do be described in terms of an effective procedure?" To now assert that there are things we know but cannot tell is not to answer the question but to shift our attention from the concept of *telling*, where until now we have tried to anchor it, to that of *knowing*. We shall see that this is a very proper and a crucially important shift, that the question of what we can get a computer to do is, in the final analysis, the question of what we can bring a computer to know. It shall preoccupy us for much of the rest of this work.

For the moment, let us recall that I have already raised that issue; earlier I said that to have a map of a city is not to know the city. Similarly, to be able to tell the rules of chess is not to know chess. The chess master knows more than he can tell. I am not saying here (although I believe this to be true) that we can never find a way to explicate the whole of his knowledge of chess; I say only that we have in this an example of knowledge that is effective even though not presently tellable. Were it true that no chess master's knowledge of chess is fully tellable, would that imply that no computer could ever play master-class chess? Not at all. We shall, as I have said, deal with such questions in what follows.

3

HOW COMPUTERS WORK

As seen from one strictly formal point of view, modern computers are simply Turing machines that operate on an alphabet consisting of the two symbols "0" and "1" and that are capable of taking on an astronomical number of states. But this is like saying that, because both bicycles and modern passenger aircraft are vehicles for transporting people, they are formally identical. The modern computer differs from the Turing machines we have been discussing both in the way it is constructed and the way it is instructed.

Many people know that a computer can compare their names as imprinted on credit cards with names somehow stored inside the computer. Yet most people believe computers are fundamentally machines that can do arithmetic on a grand scale, i.e., that they are merely very fast automatic desk calculators. Although this belief is defensible on strictly formal grounds, it is much more useful to recognize that a computer is fundamentally a symbol manipulator. Among the symbols it can manipulate are some that humans, and in a certain sense even computers, interpret as numbers. Still, most computers spend much, even most, of their time doing nonnumerical work.

To justify what I have just said, I must say something about symbols and their interpretation. And, in order to do that, I must also explain how symbols may be represented, especially inside computers.

Let us, at least for now, restrict our attention to symbols used to compose text. These are the uppercase and lowercase letters of the English alphabet, punctuation marks, and such special symbols as those used in mathematics, for example, parentheses, and addition and equality signs. The blank (or space) also counts as a distinct symbol. Were that not so, we would have a hard time writing sentences composed of individual words. Books composed of strings of only these symbols can make us laugh and cry, can tell us the history of philosophy, of an individual, or of a nation, and can instruct us in many diverse arts, including that of mathematics. In particular, they may teach us how to construct algorithms, i.e., effective procedures, and they may also give us sets of rules that constitute algorithms. Thus, however informal a notion of what information is we may appeal to, we must agree that the symbols we mean to discuss here are capable of carrying information. How are symbols represented and manipulated in computers?

Suppose that the alphabet with which we wish to concern ourselves consists of 256 distinct symbols, surely enough to include all the symbols to which I have alluded. Imagine that we have a deck of 256 cards, each of which has a distinct symbol of our alphabet printed on it, and, of course, such that there corresponds one card to each symbol. How many questions that can be answered "yes" or "no" would one have to ask, given one card randomly selected from the deck, in order to be able to decide what character is printed on that card? We can certainly make the decision by asking at most 256 questions. We can somehow order the symbols and begin by asking if it is the first in our ordering, e.g., "Is it an uppercase A?" If the answer is "no," then we ask if it is the second, and so on. But if our ordering is known both to ourselves and to our respondent, there is a much more economical way of organizing our questioning. We ask whether the character we are seeking is in the first half of the set. Whatever the answer, we will have isolated a set of 128 characters among which the character we seek resides. We again ask whether it is in the first half of that smaller set, and so on. Proceeding in this way, we are bound to discover what character is printed on the selected card by asking exactly eight questions. We could have recorded the answers we received to our questions by writing "1" whenever the answer was "yes" and "0" whenever it was "no." That record would then consist of eight so-called *bits* each of which is either "1" or "0". (When speaking in terms of decimal notation for numbers, we refer to the numbers 0, 1, . . ., 9 as *digits*. But our notation permits us only the two symbols "0" and "1"; we refer to them as *bits*.) This eight-bit string is then an unambiguous representation of the character we were seeking. Moreover, each character of the whole set has a unique eight-bit representation within the same ordering.

There do, in fact, exist widely agreed upon conventions for ordering just such a set of characters and for their individual encodings. (That these conventions are not universally agreed to need not concern us here, at least not for the moment.) In recent years the specific coding scheme used in computers manufactured by the IBM company has become very nearly a worldwide industry standard. Within that convention an eight-bit string representing a character (of a 256-character alphabet) is called a *byte* and a chain of four bytes a *word*.

We have seen that any text can be represented as a string of 1's and 0's. To do any useful work on information encoded as bit strings, we must be able to manipulate them in some orderly way, i.e., to play games with them. We now know the pieces of the games we may wish to play. All that remains is to state rules. But before we come to that, let me say a few words about the electrical representation and manipulation of bit strings.

We may say about any wire that an electric current is flowing in it or not. Consider a wire connected to a source of electric power and a suitably connected switch. When the switch is closed, current flows through the wire, otherwise not. Suppose that the switch is connected to a mechanism that opens and closes it regularly; say, the switch is closed for one second, then open for one second, and so on. We may then speak of the flow of electricity on the wire as a pulse train (see Figure 3.1.) An ordinary electric doorbell is a pulse generator. When power is supplied to it, i.e., when the bell button is pushed, a switch is closed. Current flowing through a wire causes the bell's hammer to move, opening the switch. The switch is then again closed, and so on.

A modern computer is, of course, fundamentally an electrical device, just as an electric doorbell is. When we push a bell button, we think of the bell as being "on," even though it, in a sense, turns itself on and off while the button remains depressed. An operating computer may be thought of as being similarly on and turned on and off by a train of pulses such as we have discussed. Conceptually, then, a computer's time is divided into two kinds of intervals, a quiescent interval during which it is, in a sense, "off," and an active interval during which anything that is to happen must happen. In effect, the regular pulse train we have discussed acts as a clock. A state of "no current" on the wire carrying that train signals the quiescent period, and current on it signals the active period.

To conceptualize what goes on inside a computer, think of a railway map, representing the entire rail network of a continent, in which the actual rail lines are represented by wires. Each railway station is represented by a pair of little neon bulbs. When we look at this network of wires and bulbs during a quiescent interval, we notice that some bulbs are lit and others are dark. During the next active period some of the lit bulbs go off, some dark ones go on, and some remain as they were. That's all. Then there is another quiescent period. There was no flickering of bulbs during the active interval. Each bulb either remained as it was or changed its state exactly once, i.e., turned off if it was on or turned on if it was off.

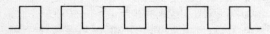

Figure 3.1. A pulse train of square waves.

The device we have conceptualized as a pair of neon bulbs is an electronic circuit consisting of two identical components. Each component is capable of circulating an electric current indefinitely, i.e., of in effect *holding* a pulse. However, only one of the pair may hold a current at any one time. Two wires lead into the device, one to each component. When, during an active period, a wire transmits a pulse to a component that is not then circulating a current, a current is induced in it and the current circulating in the other half of the device is shut off. This device is thus able to flip and flop between two states; either one of its halves is "on" and the other "off" or vice versa. It is therefore called a *flip-flop*. Each of its components also has a wire emanating from it, and each wire will, of course, carry a current, i.e., a pulse, during an active period when the half of the flip-flop corresponding to it is "on." The function of a flip-flop in a computer circuit is to "remember" on which of its two sides a pulse last impinged. It is a one-bit information-storage device.

We now have another way of saying what goes on inside a computer during an active period: Many flip-flops change state. But the function of a computer is to manipulate information, not merely to transmit it from place to place. And information manipulation is, as we have already observed, fundamentally a matter of transformation. Now any single wire leading from a side of one flip-flop to that of another can carry at most one pulse during a single active interval. Therefore, whatever transformations are to be achieved during an active interval must be results of electrical operations, not on streams of pulses following one another in time, but on a line of pulses advancing in parallel.

Suppose there were an enormous telephone network in which each telephone is permanently connected to a number of other telephones; there are no sets with dials. All subscribers constantly watch the same channel on television, and whenever a commercial, i.e., an active interval, begins, they all rush to their telephones and shout either "one" or "zero," depending on what is written on a notepad attached to their apparatus. They also listen for what is being shouted at them, and write either "one" or "zero" on their pads, depending on what they hear. A telephone may be a transmitter to several receivers at once, and it may also be a receiver for more than one transmitter. Yet the signal reaching each receiver must be an unambiguous "one" or "zero." There must therefore be operators (actually, electronic devices), placed along the wires connecting receivers to one another, whose function is to compose a single signal, "one" or "zero," from possibly many incoming signals. These devices are called gates.

Let us describe three different kinds of gates, each with a distinct function. The simplest is one that, when it receives a "zero" transmits a "one" and vice versa. This is the NOT gate. Its function is described by the formulas

$$NOT(0) = 1,$$
$$NOT(1) = 0.$$

A common schematic representation for it is shown in Figure 3.2. It has one input, here labeled A, and one output, here labeled B.

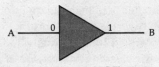

Figure 3.2.   A NOT gate.

The AND gate has two inputs and one output. It transmits "one" if and only if its two inputs are both "one"; otherwise it transmits "zero." Its function is described by the formulas

$$AND(0,0) = 0,$$
$$AND(0,1) = 0,$$
$$AND(1,0) = 0,$$
$$AND(1,1) = 1.$$

Its common schematic representation is shown in Figure 3.3.

The OR gate also has two inputs and one output. It transmits

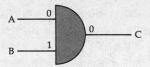

Figure 3.3. An AND gate.

"one" whenever either or both of its inputs are "one"; otherwise it transmits "zero." Its formulas are

$$OR(0,0) = 0,$$
$$OR(0,1) = 1,$$
$$OR(1,0) = 1,$$
$$OR(1,1) = 1.$$

Its common schematic representation is shown in Figure 3.4.

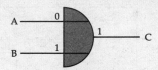

Figure 3.4. An OR gate.

The highly motivated reader may wish to trace pulses through the system of gates shown in Figure 3.5, which represents a circuit whose components are AND, OR, and NOT gates and whose function is to arithmetically add two binary digits.

(Binary addition is like decimal addition, only much simpler. The decimal sum of 2 and 3 is 5. The decimal sum of 7 and 8 is also 5 but there is also a "carry" of 1, which is added to whatever the sum of the next column to the left is. In binary arithmetic, there are only two digits, 0 and 1. The addition rules are very simple, namely,

$$0 + 0 = 0,$$
$$0 + 1 = 1,$$
$$1 + 0 = 1,$$
$$1 + 1 = 0 \text{ and carry } 1.$$

80   *Chapter 3*

The one-bit binary adder shown therefore has three inputs, *A, B,* and *C,* and two outputs, *S* and *D. A* and *B* represent the two bits to be added, and *C* any carry that may have been produced by a similar adder to the right, so to say, of the one shown here. *S* is the sum produced, and *D* the carry.)

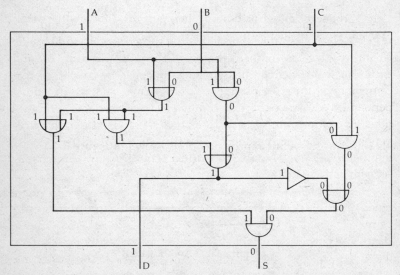

Figure 3.5. A one-bit adder with carry-in and carry-out. (Figures 3.2–3.5 from D. C. Evans, "Computer Logic and Memory," Copyright © 1966 by Scientific American, Inc. All rights reserved.)

It is really not necessary for the reader to trace pulses going through this adder. The important thing is to understand that combinations of the three simple gates we have described are capable of realizing transformation rules on information. Notice also that we enclosed the whole circuit in a box. This box has three inputs (*A, B,* and *C*) and two outputs (*S* and *D*), and is itself a unit. We have combined simple components to make a more complex component. We could now combine, say, 32 of these adders and form a 32-bit adder. And that would then be a single component. Both the construction and the instruction of computers is just such a process of making bigger things out of smaller things.

Let us return for a moment to our image of interconnected telephones, but this time with the realization that gates intervene in conversations (if we may call them that) among subscribers. We can now imagine the three subscribers *A, B,* and *C* picking up their telephones at the appropriate time and shouting "1," "0," and "0," respectively. If they are connected to the subscribers *S* and *D* by the circuit we have shown, the *S* will hear "1" and *D* will hear "0," and each will write what he hears on his notepad. Of course, in real computers all these "subscribers" are flip-flops, and the broadcast periods are extremely short. In computers, then, results of computations, i.e., of information transformations, performed during active periods are stored in flip-flops. There they survive quiescent periods in order to become available for further transformation during subsequent active periods.

I have already suggested that one-bit adders may be combined to form, say, 32-bit adders. The inputs to such an adder would be two sets of 32 flip-flops each—the carries are internal to a multistage adder—and the output again a set of 32 flip-flops. Each of these sets may be said to contain a 32-bit binary number during quiescent periods. Each would of course have to be an ordered set, i.e., one in which there is a first bit, a second bit, and so on. Such a set of flip-flops is called a *register.* It is another example of an aggregation of more elementary components. But may we say about any *n*-bit register that it contains an *n*-bit number? No, at least not unconditionally; for what interpretation may be placed on a bit string residing in a register depends on what components are wired to that register, hence what operations may be performed on its contents. If a particular pair of registers is connected only to components that perform arithmetic operations, i.e., if the computer treats the information stored in them only as numbers, then they are numbers—at least while they are being manipulated within the computer. To appreciate that the symbols that occur in natural language are subject to similar constraints, one need only consider this very sentence, in which the word "one" is a number in one place, a word in another, and an uninterpreted character string in a third.

An operating computer is engaged in playing an elaborate and very complicated game. After each quiescent interval it makes a move. The contents of many registers are transported to other registers much as chess pieces are moved over the chess board. But in the computer a great many pieces, bits, are moved at once and, what is most important, individual bits and sets of bits are transformed while on their journey from place to place.

I have already mentioned the fact that in many computers sets of eight bits are aggregated into bytes. An eight-bit register is a physical embodiment of this conceptual aggregation. Computers contain many registers that are connected to one another by combinatorial gating networks of the kind I have described. The entire set of these registers and of the logical networks that unite them constitutes, in a sense, the computer. This set is the machine, often called the computer's central processing unit (CPU), that actually performs logical symbol manipulation during each of the computer's active periods. The total configuration of states of its individual flip-flops constitutes the state of the computer in a sense strongly analogous to what I meant when earlier I spoke of the state of a Turing machine.

A Turing machine of the kind we discussed earlier gets the information on which it is to operate from a tape. It must read and write this tape sequentially, one symbol at a time. A modern computer, on the other hand, stores much of the information it manipulates internally. The computer's internal-storage device consists of a very large array of eight-bit registers, possibly a million or more of them. These are arranged in a definite order, somewhat like the mailboxes in the lobby of a large apartment house. The registers are numbered serially, beginning with 0, 1, and so on. A register's number is called its *address*. Since each register has a unique address, one can speak of a register's address quite independently of its contents, i.e., the state of its eight flip-flops, and vice versa. Now imagine an array of apartmenthouse mailboxes that is equipped with exactly one combination lock. In order to take anything from a specific mailbox or put anything into it, one must first set that mailbox's address into the lock. The computer's store has just such a device. It is, of course, a register. (A 20-bit register would be sufficient to address 1,048,576 boxes.) It is useful to think of this address register as part of the computer's CPU, even though it is connected to both the computer's store and its CPU. But to see it as part of the latter helps to visualize that its contents are themselves machine-manipulable. They may, for example, be intermediate results in some long chain of arithmetic computation and may be used as, in effect, ordinal numbers.

The addressability of a modern computer's store is one of its most important properties. To appreciate the enormous difference addressability makes just to the way one searches a store, consider the following problem. A certain town has fewer than 10,000 telephone subscribers. The telephone directory for that town lists all subscribers alphabetically and, of course, gives their respective telephone numbers. But it also lists each subscriber's serial position in the directory. For example:

1 Aaban, John 369-6244

.

.

.

1423 Jones, William 369-0043

We wish to know whether the last four digits of any subscriber's telephone number are the serial number of any other subscriber whose telephone number similarly corresponds to the first subscriber's serial number. Given the listing shown for Jones above, for example, is there a listing of, say,

43 Baker, Max 369-1423

which would meet these conditions?

A simple way to solve the problem is to look at each listing beginning with the first, and see if it and the listing with the serial number corresponding to the last four digits of its telephone number constitute a pair of the kind we are seeking. The answer is "yes" if we find one such pair, and "no" if, after inspecting the whole set, we find no such pair. The worst case we could have encountered is that in which no pairs exist. Then we would have looked at every listing at most twice. But suppose the telephone directory were recorded on a tape of the kind required by the Turing machines we have described. Then, even apart from tape motion associated with book-keeping functions, every listing would need to be scanned many more times than twice. The very first listing in our example would require that we look at it, at the 6242 listings between it and the 6244th listing, and at the 6244th listing itself. In classic Turing machines such a tape would not be merely passed over between relevant listings, but each listing would actually have to be inspected and interpreted. The same principle accounts for the anger some people feel when told they are mentioned in a large book that unfortunately has no index. They must then face the prospect of having to read the whole book.

This telephone directory example illustrates not so much that addressability increases the efficiency of searches, but that it can, under some circumstances, help avoid the need to search at all. For had we specified a particular subscriber and asked whether he is paired with another in the way we indicated, that question could have been answered directly and without searching through irrelevant data. Is Mr. Aaban so paired, for example? To find out we look directly at subscriber number 6244. If the last four digits of his telephone number are "0001," then yes; otherwise no.

With this in mind, let us look back at the quintuples that define some Turing machine T and thus constitute a program for a universal Turing machine that is to imitate T. Recall that the general form of such a quintuple is

(present state, present symbol, next state, new symbol, direction).

In general, a Turing machine in a certain state, say, 19, and scanning a certain symbol, say, "1," must read through all quintuples in its program, constantly asking, so to say, whether the particular quintuple (rule) it is currently reading is the one that applies to state 19.

When it finds the one appropriate to its state and to the symbol it is then scanning, it rewrites that symbol, moves its tape, and (possibly) changes state. Then the search begins again. If, however, the quintuples were recorded in an addressable store, then the search for the appropriate quintuple could be avoided or at least reduced in length. Suppose, for example, that all the quintuples associated with a particular state required, say, 100 bytes of storage space and that the first one is stored beginning in register 1000. Then the set corresponding to state 19 would be stored beginning in the register numbered $1000 + (19 - 1) \times 100$, i.e., register 2800. The whole notion of state is thus transformed into that of address, at least in this context.

In real computers, data too are recorded in the computer's addressable store. This innovation allows us to eliminate as well the restriction that only a datum immediately adjacent to one presently being scanned may be immediately accessible. To begin to see how addressability is used in the composition of real computer programs, let us look at a small but realistic problem.

We want to compute the square roots of numbers. (That is, given a number, say, 25, we want to know what number, when multiplied by itself, will produce the given number. The square root of 25 is 5, because $5 \times 5 = 25$.) We assume we have a faithful (human) servant who will tirelessly obey every instruction we give him. We know an algorithm for computing square roots of positive numbers. Given the number n, we compute its square-root as follows:

1. First we make a guess, always the same one, namely, 1.

2. We then arrive at a better guess by:
   (a) multiplying the old guess by itself;
   (b) adding the given number n to that product; and
   (c) dividing that sum by twice the old guess.
   (No less an authority than Isaac Newton proved that this
       computation always yields a better guess, unless, of course,
       the previous guess was already the correct solution.)

3. If the difference between the old guess and the better guess is small enough for our purposes, then we accept the new guess. Otherwise we compute a still better guess.

We justify this procedure as follows.

Suppose we wish to find the square root of 25 and we take our initial guess to be, say, 4. We know that this is too small, since 4 × 4 = 16. If we divide the number originally given, that is, 25, by our guess, then the quotient will be larger than the result we are seeking, which is 5 in this example. That quotient and the guess we have made therefore bracket the result we seek. If we then take the average of the two, we will get another guess, moreover, one that is closer than the guess on which it was based. We may iterate on this formula until we get a result as close to the correct one as we please.

Our servant, however, is not terribly bright. He has worked with tax forms that require one to calculate this-and-that and to write it on line so-and-so or in box such-and-such. We have therefore made up the worksheet shown in Table 3.1, namely, one based on such procedures. Our servant works in a little cubicle that has an input slot and an output slot. As soon as a slip of paper with a number written on it appears in the input slot, he begins to work furiously. When he finishes he writes his result on a paper and puts it in the output slot. His only initial instruction is to start by obeying the instruction on line 101 of the worksheet and, unless he encounters an instruction to the contrary on the worksheet itself, to continue obeying the instructions in the sequence in which they are written. He would, by the way, be well-advised to write only in pencil and to have a good eraser at hand; he must use what little space is given him for writing on the worksheet over and over again. (The reader should not attempt to carry out an example computation on the worksheet to the bitter end. He may, however, profit from carrying it out sufficiently far to generate, say, two better guesses for the square root of 25.)

Notice the important role box A plays. It is in effect a register that contains intermediate results of computations. Notice also that no individual instruction refers to more than one line number. Instructions that have this property are called *single-address instructions.* We can perform, say, additions—which, of course, require two operands—by first storing one operand in box A, then adding the other operand to the contents of box A, and leaving the sum again in box A. Lines 121, 123, and 124 serve as temporary storage registers. Finally, notice that the instructions and the storage of intermediate results are so organized that a worksheet once used may be used again. On second and subsequent uses, box A and lines 121, 123, and 124 will contain numbers irrelevant to the new task. But these numbers will not interfere with the newly started computation. In

Table 3.1. *Directions for computing square roots.*

101	Get number from input slot and write it on line 121.
102	Copy contents of line 122 into box A.
103	Copy contents of box A onto line 123.
104	Copy contents of line 123 into box A.
105	Add contents of line 123 to contents of box A, and write result in box A.
106	Copy contents of box A onto line 124.
107	Copy contents of line 123 into box A.
108	Multiply contents of box A by contents of line 123, and write result in box A.
109	Add contents of box A to contents of line 121, and write result in box A.
110	Divide contents of box A by contents of line 124, and write result in box A.
111	Copy contents of box A onto line 124.
112	Subtract contents of line 123 from contents of box A, and write result in box A.
113	Copy the absolute value of box A into box A.
114	Subtract contents of line 125 from contents of box A, and write result in box A.
115	If contents of box A are greater than zero, begin work with line 118.
116	Put contents of line 124 in output slot.
117	Stop.
118	Copy contents of line 124 into box A.
119	Copy contents of box A onto line 123.
120	Begin working with line 105.
121	0
122	1.0
123	0
124	0
125	.001

Box A

this we followed a quite universally accepted programming practice: whereas many public washrooms display a sign urging users to leave the room as they found it, we adopt just the opposite convention. We say "Put the room in the condition you wish it to be in before you begin serious work." We always store the "old guess," for example, in line 123. After the worksheet has been used, that line will contain the last "old guess" of the last completed computation. But a fresh computation, for which the last stored "old guess" is totally irrelevant, begins by copying "1," our standard first guess, into line 123 (see lines 102 and 103).

For a reason that appears to be nowhere recorded, a computational procedure of the kind we are here discussing is called a *routine.* We speak of beginning such a computation as entering the routine, and when we have completed the last step, we say we leave the routine. But small routines such as the one illustrated here are seldom of much use in and of themselves; how many times does one really need to know the square root of a number outside of a context established by some larger computational task? Such a larger task may be, say, the computation of the roots of a quadratic equation, i.e., of an equation of the form

$$ax^2 + bx + c = 0.$$

The desired roots are given by the formula

$$x = \frac{-b \pm \sqrt{b^2 - 4ac}}{2a}.$$

Here then is a problem—i.e., to find the roots of a given quadratic equation—which involves the computation of the square root of some number as a subproblem. A worksheet very similar to Table 3.1 can be designed that would direct a human computer in the solution of the over-all problem. But, since we already have a worksheet that tells how to compute square roots, we would be well-advised to tell the person working on the larger problem to use the already prepared worksheet to help him solve the subproblem when he comes to it, i.e., after he has computed the quantity $(b^2 - 4ac)$. The square-root routine then becomes a *subroutine* of the larger routine. Of course, that larger routine may again be a subroutine of a still larger routine, and so on. Again I emphasize that much of computers and computation has to do with building larger hierarchical structures out of smaller ones. It is precisely because the square-root routine we have exhibited may play the role of a subroutine in a larger procedure that we have taken care that it initializes itself, that is, puts itself in good order, before each use.

The perceptive reader will have noticed that I have glossed over a management problem that will surely come to haunt us if our computing servants are really as simple, as literal, and as unimaginative in the way they interpret the instructions we give them as I said they were. We have assumed, and we shall stick with the assumption, that a worker knows how to add a number read from a certain line to the contents of box A, how to leave the sum in box A, how to tell whether a number recorded in box A is greater than zero or not, and so on—that, in other words, he knows enough to be able to interpret and to obey all the instructions on the square-root worksheet. But how do we instruct someone similarly trained but working on larger problems to subcontract, so to say, his subproblems?

We can avoid having to confront that question by simply assuming that the worker knows how to compute, say, square roots just as we assume he knows how to add and subtract. But are we also to assume he already knows how to compute the roots of quadratic equations? If we do not stop making assumptions of this kind we will assume the need for computers away altogether, for then we will have postulated that, in order to compose a worksheet corresponding to any computation whatever, all we have to write is, in effect, "do it." We may be entitled to believe that people can add, subtract, and so on. If we want them to perform more complicated symbol-manipulation tasks, however, we will have to describe appropriate procedures to them in terms of things they already understand. On the other hand, we don't want to have to describe an often-used procedure over and over again. Every routine is potentially a subroutine of some larger routine. Therefore, having once written a particular routine, we would want it to become part of a subroutine library from where it can be called whenever it is needed.

Every library has some convention for calling volumes from the stacks; a reader fills out a request card, hands it to the librarian, etc. But our situation is a little more demanding than that of an ordinary lending library. We want our workers not only to see the worksheet corresponding to the called-for computational task, but also to, in effect, farm out the work itself. For example, suppose a cook has brought a recipe to a certain point and then wishes a specialist to perform the function he is famous for on ingredients the cook has already prepared. The cook may rest while the specialist works, and may continue his labor only after the specialist returns the desired concoction to him. Assuming there are many such specialists, how does the cook turn control of the kitchen over to just the right specialist, how does he make sure the specialist works on the right ingredients, how does he finally regain control over his kitchen, and where does he find the results of the specialist's labors? One way to do all these things is as follows.

> The cook leaves all the ingredients the specialist is to work on in sequentially numbered cupboards, say, the first in 119, the next in 120, and so on. In the next cupboard after the one containing the last ingredient, he writes the number of the apartment he will be visiting while the specialist works. He leaves the number of the cupboard in which the first ingredient is stored on the stove. Finally, he rings the bell of the specialist's apartment, whose address he, of course, knows.
>
> The specialist knows the cook's conventions very well. When he hears the bell ring, he goes to the kitchen, finds on the stove the number of the cupboard containing the first ingredient, and begins doing his tricks. When he is finished, he leaves his concoction on the stove, rings the bell of the apartment whose address he found in the first cupboard after those that contained ingredients, and leaves.

In order for this scheme to work, it is necessary only that the cook and every specialist who may be called on know the conventions, that the cook know exactly how many ingredients the particular specialist he has called needs (he wouldn't want the literal-minded man to boil the note containing the cook's address along with the other ingredients of the stew), and that the cook know the address of the particular specialist he wishes to call.

We can invoke virtually identical conventions for calling subroutines in our worksheet format. Consider, for example, the following fragment of a worksheet. It is part of a larger routine. At the point we take it up, a number whose square root is to be computed is in box A. This fragment calls the square-root subroutine and also assures that control will return to the main routine.

508. Copy contents of line 680 into box A.
(Note: Line 680 contains the number whose square root is to be computed.)

509. Copy contents of box A onto line 512.
(Note: Line 512 is the "cupboard" in which the first—in this example, the only—"ingredient" is stored.)

510. Copy the number "512" into box A.
(Note: Box A plays the role of the store. The specialist has now been told where to begin to look for ingredients.)

511. Begin working with line 98.
(Note: The square-root subroutine begins on line 98.)

512. (Note: This line serves as storage for the number whose square root is to be taken.)

513. 514.
(Note: This line contains the address of the line to which the specialist is to return control after he finishes his work.)

514. (Next instruction in larger routine comes here.)
(Note: When this instruction is encountered, the square root of the number stored in line 680 is in box A.)

The specialist's worksheet has already been displayed, in Table 3.1. However, we must modify it to take account of the conventions we are discussing. The original worksheet tells the worker to get the number on which he is to operate from an "input slot." Now, of course, the number of the line containing that datum is in box A. We therefore write:

98. Copy the contents of box A onto line 126.

99. Copy the contents of the line whose number is contained in line 126 onto line 121.
(Note: This is called indirect addressing. The routine was originally set up to operate on a number stored in line 121. In this way we can leave the routine undisturbed.)

100. Add "1" to the contents of box A, and leave the sum in box A.
(Note: The contents of box A were undisturbed by the execution

of the instructions in lines 98 and 99. Hence box A now contains "513," which is the line number of the calling routine which contains the return address; line 513 in this example contains "514.")

101. Copy contents of box A into the address portion of line 117.
(Note: Line 117 was a stop instruction but we will replace it as shown below. Since all instructions are single-address instructions, the term "address portion" is unambiguous when applied to instructions.)

We now rewrite lines 116 and 117 as follows.

116. Copy contents of line 124 into box A.
(Note: We are leaving the "concoction," i.e., the square root of the number given us, in box A.)

117. Begin working with line 0.
(Note: The "0" in this line will, of course, be replaced whenever the instruction in line 101 is obeyed, in other words, each time this subroutine is invoked.)

My aim has been to illustrate one of the many possible ways to call a subroutine. Although the specific way illustrated here works, it is not as efficient as some others, and should not be considered standard. What matters is the principle that, once a routine is written, it can be made a component of a larger routine. That larger routine can then again be treated as a subroutine of a still larger routine, and so on. One reason this kind of hierarchical building of computing (i.e., symbol-manipulating) structures is important is that it relieves the designer of the larger, higher-level system from having to know precisely how (i.e., by what algorithm) the lower-level subroutine does its work. He needs to know how to get at it, what its calling conventions are, and what the functional relation of its output is to its input. A subroutine is therefore rather like a sometimes complex legal instrument, say, a building lease: one fills in all the blank spaces, signs it, and files it. Usually each signatory believes correctly that exactly the legal consequences he had in mind have been entrained. But sometimes one gets into trouble when relying on prepared forms whose "intentions" one thinks one understands and whose fine print one fails to read. I shall have more to say about such things later.

A human worker whose job it is to compose the kind of worksheets I have described would soon tire of having to write so much. He would soon invent abbreviations. No essential information is lost if, for example, the line

(103. Copy contents of line 123 into box A)

of the original worksheet (Table 3.1) is abbreviated by

(103. GET 123)

or if the line

(105. Add contents of line 123 to contents of box A, and write result in box A)

is abbreviated by

(105. ADD 123).

We can invent similar abbreviations (in effect, verbs) for each of the operations mentioned in the original worksheet, and use them to encode the whole procedure as in Table 3.2.

The procedure so encoded is still readable by people. What is more important to us at the moment, however, is that such a code is eminently readable by a computer. Not that the original worksheet written in English is not. After all, we have seen that any text can be reduced to a string of 0's and 1's. But the code shown in Table 3.2, shorn of its commentary, has a rigid format and is therefore easy to decode. Each line is a command to the computer, and each such command consists of two components: an operation to be performed, e.g., ADD, and the address of an operand. Box A is no longer mentioned explicitly. It is understood that box A always contains the implicit or unmentioned operand when one is needed, and that the result of any operation, e.g., the sum produced by an addition, is stored there. But from a computer's, as opposed to a human's, point of view, this code is still much too longwinded. The

Table 3.2.   *An encoding of a square-root routine.*

LINE	COMMAND	COMMENT
101	INP 121	Get number from input and put in 121.
102	GET 122	Get initial guess.
103	STO 123	Store as "old guess" in 123.
104	GET 123	Get "old guess."
→ 105	ADD 123	Double it.
106	STO 124	Store "twice old guess" in 124.
107	GET 123	Get "old guess."
108	MPY 123	Multiply it by itself.
109	ADD 121	Add given number to that.
110	DIV 124	Divide result by "twice old guess."
111	STO 124	Store "new guess" in 124.
112	SUB 123	Subtract "old guess" from "new guess."
113	ABS	Make that result positive.
114	SUB 125	Subtract tolerance from that.
115	JGZ 118	If that is greater than zero, skip next two steps.
116	OUT 124	Put out "new guess."
117	STP	Stop.
118	GET 124	Get "new guess."
119	STO 123	Put it in place of "old guess."
120	JMP 105	Start "loop" again.
121	0	Place for number to be worked on.
122	1.0	Place for initial "old guess."
123	0	Place for "old guess."
124	0	Place for "new guess."
125	.001	Place for tolerance.

operation-code portion of a command, e.g., ADD, can be replaced by a single byte-sized character. (Recall that there are 256 distinct such characters.) The only restriction such a convention then imposes is that we cannot appeal to more than 256 so-called built-in operations, i.e., operations that require no further explanation to the worker. Such operations are called *primitives.*

To continue the conversion of this example procedure to machine code, suppose that the byte-length code for ADD is "00110101." Then, if the length of a command as stored in a particular computer is four bytes, i.e., 32 bits, the actual computer code for line 105 of our worksheet would be

00110101000000000000000001111011.

The whole code for our routine would be an array of 25 such lines.

We have illustrated how a computational task may be organized as a procedure executable by humans and how such a procedure may be translated into a notation that, although cryptic, is easily managable by a computer. But we have cheated just a little. Our advice to our faithful computing servant was to obey the instructions on the worksheet in the sequence in which they are written unless a specific instruction tells him to do otherwise. We had to assume, of course, that he understood and knew how to execute the primitive operations called for on the worksheet. But we also tacitly assumed that he could remember his place on the worksheet even while distracted during, say, the performance of a long division. Actually, maintenance of the flow of control in a program is a task quite different and separable from that of executing primitive operations. We should really add another box to the worksheet, namely, a box P (for *program counter*) which at all times contains the line number (i.e., the address) of the instruction then being obeyed. We should then appoint a supervisor whose job is to tell the worker on what line he is to find his instruction, namely, the line whose address is stored in box P. When the worker tells the supervisor that the assigned instruction has been obeyed, the supervisor adds "1" to the contents of box P, stores the new count there, and gives the worker his next assignment. The effect of an instruction that disturbs the normal sequential flow of control is, of course, to replace the contents of box P with the address of the next instruction that is to be obeyed. For example, the effect of obeying the line

(120. Begin working with line 105)

is that the worker copies the number "105" into box P and immediately goes about obeying the instruction written on line 105 without reporting to the supervisor that he has finished with one instruction.

In real computers, work is subdivided much further still.

There are components for accomplishing each of the built-in or primitive operations of which the computer is capable. These, although they may share some circuitry, are essentially separate. There are still more components that manage access to the computer's store. Many computers even have small special-purpose sub-computers whose sole task is to supervise the transfer of information between the outside world and the main computer. I cannot discuss such details here without being led much too far from my main concerns. I did, however, single out the idea of flow of control of a program, because it is a really important concept. Notice that two lines are drawn to the side of the program shown in Table 3.2. These illustrate the flow of control in that program, or, to put it another way, the structure of that program. What makes even this little program at all interesting is that it involves a *conditional branch*. Whether or not the program, while running, goes through another iteration is determined by the outcome of the instruction of line 115. If, when that instruction is encountered, box A contains a number greater than zero, then another iteration is entrained. Otherwise, the program stops after another few steps.

The ability of computers to execute conditional-branch instructions—i.e., to modify the flow of control of their programs as a function of the outcome of tests on intermediate results of their own computations—is one of their most crucial properties, for every effective procedure can be reduced to a series of nothing but commands (i.e., statements of the form "do this" and "do that") interlaced with conditional-branch instructions. Moreover, only binary branching instructions (i.e., instructions of the form "if such-and-such is true, do this; otherwise do that") are needed. If the decision whether such-and-such is true or not itself involves complex procedures, these too can be cast into the framework of commands and binary (i.e., two-way) branching instructions.

I have now drawn a fairly accurate outline—albeit only an outline—of what a program is and of how a computer executes a program. In the course of the argument, we reduced a very wordy worksheet to a highly compact code, and then reduced that code still further, to a rigidly formatted bit string. We assumed all along that none of these reductions were accompanied by the loss of any essential information. We hope, in other words, that the string of 0's and 1's which is the final product of all our transformations means the same thing as our original worksheet. To test whether or not it does, we must know what each means separately.

What does the worksheet we have made up mean? To an illiterate it may mean an opportunity to build a paper airplane; and to a mother who believes her ten-year-old child composed it, that her child is a budding genius. But in our context, we see it as a set of instructions. Hence its meaning to us is the action someone who understands those instructions takes when he obeys them. The meaning of the computer code must then be the action a computer takes when it interprets that code as instructions and obeys them. We are confident we know what a person following the instructions written on the worksheet will do, because they are written in a language he and we share. What a computer will do, given the code we prepared, is entirely a function of its design. That design determines the meaning of any string of 0's and 1's given a computer. If the particular bit-string we have developed is to mean the same thing to a computer as what our worksheet means to a human calculator, then the computer must have components functionally equivalent to each of the operations we ask our human calculator to perform and to every resource we make available to him.

I have, of course, been very careful to structure this discourse about workers, supervisors, A boxes, P boxes, and so on, in such a way that these resources and functions would be analogous to components and functions of real computers. In fact, real computers do have A boxes, namely, registers called accumulators, and P boxes, namely, registers called program counters. Real computers do store their instructions in sequentially numbered registers, and they do manage the flow of control in roughly the way I have pictured. Modern computers are much more complex, both in their structure and in their functions, than my simple sketch of one kind of computer reveals, to be sure. But the fundamental operating principles I have described do provide a starting point for understanding even them.

When speaking of Turing machines, I asserted that a language is a game played with a certain alphabet and governed by a set of transformation rules, and that a computer is an embodiment of those rules. That assertion applies with equal force to modern digital computers. Procedures can, as we have seen, be encoded as bit strings whose format is dictated by certain design features of the computer for which they are intended. The primitive vocabulary that a programmer may employ is determined by the operations built into the computer. For example, a particular computer may have the square-root function built in as one of its primitive operations. The language corresponding to that computer would contain an operation code corresponding to that function, in effect, a verb meaning "take square root." In the language determined by some other computer, one that does not have "take square root" as a primitive operation, the square-root function can be realized only by a subroutine, in effect, a verb clause composed of more elementary terms. I wish to emphasize again that every computer determines a language, its machine language. The alphabet of the machine languages of all modern computers is the set consisting of the two symbols "0" and "1." But their vocabularies and their transformation rules differ widely.

Do computer programmers then work in the language of the machine they are instructing? I.e., do they actually compose complex procedures in terms of long bit strings? No. We have seen that it is straightforward and easy to translate the expression

(ADD 123)

into the bit string

00110101000000000000000001111011.

A computer is a superb symbol manipulator. It is relatively easy to design a procedure that will convert a program, such as the one in Table 3.2, into its corresponding bit string. Indeed, such procedures can also be made to handle mundane, but often far from trivially easy, bookkeeping tasks, such as the detailed assignments of storage locations. Programs that do this sort of work, i.e., that convert programs written in the kind of notation shown in Table 3.2 into machine language, are called *assemblers*. The language from which they convert, i.e., the language they translate, is called an *assembly language*.

Programs written in assembly language are, of course, much easier for people to read than programs in machine language. Since assemblers are themselves fairly complex programs, they too are written in assembly language, namely, the language they themselves translate. There is no paradox in this. An assembler is a procedural embodiment of well-determined rules that tell how to transform expressions composed on a certain alphabet into expressions composed on a much smaller alphabet. It therefore defines a language, its assembly language. Since it is designed to translate any legal assembly-language text into machine language, there can be no mystery about its ability to translate the text that constitutes its own definition as well.

I said earlier that a program transforms a computer into another computer. A general-purpose computer loaded with nothing but a square-root program, for example, has been transformed into a special-purpose computer capable of computing only square roots. An assembler, then, also transforms the computer into which it is loaded. The transformation it induces has important consequences: a programmer who instructs a computer using only an assembler need never learn the language determined by the computer itself, i.e., its machine language. In an important sense, he never sees the machine he is actually addressing; he sees and works with a symbolic artifact that, for him, *is* the machine.

It did not take long for programmers to realize that the symbol-manipulating power of the computer could be employed to translate still "higher level" languages. Consider, for example, the following fragment of assembly language text:

> GET a.
> ADD b.
> STO c.

With even as little as we covered in the preceding pages, we can see that that fragment must be an encoding of the algebraic expression

$$c = a + b.$$

Clearly a very simple procedure will do to translate the latter expression into the assembly-language fragment shown. Now, without fretting much over details, it may be said correctly that the following is an encoding of the square-root procedure with which we have already had so much experience.

```
Procedure SQRT(n):
    Let tolerance = .001;
    Let oldguess = 0;
    Let newguess = 1;
    While |oldguess-newguess| > tolerance do:
        oldguess = newguess;
        newguess = (oldguess**2 + n)/2*oldguess;
    End;
    Result = newguess;
End of procedure SQRT.
```

To translate that procedure into assembly language is relatively simple. We did essentially just that in some preceding pages. It is far more difficult however, to write a procedure to translate into assembly language any program written in the language of which the above is a sample text. It may appear at first glance that this difficulty results from the obvious complexity of the language as compared to the stark simplicity of assembly languages. It is true that the translator must be able to handle a very great variety of syntactic forms, many of which, especially as they occur in combinations, are the source of truly difficult technical problems. Still, these are not the hardest problems.

We must ask what it means to say that the procedure we have shown is a "sample text" of some language. Of what language? The problem of first priority, and by far the hardest one, is to design the "higher level" language at all. The formation and transformation rules of a language (as of a game) are, after all, as much a system of constraints, of prohibitions, as of permissions. The very fact that the basic building blocks used to compose assembly-language programs are so utterly elementary allows the programmer an enormous number of degrees of freedom. An assembly language for a particular machine is virtually a mnemonic transliteration of its machine language. The designer of an assembly language is therefore hardly confronted with questions of meaning. But the designer of a higher-level language must decide the kinds and number of degrees of freedom to be allowed to authors writing programs in that language.

In a very real sense, every freedom an author exercises in making a choice is one he has usurped from the programmer who is to use his product. Consider a very simple example, namely, the expression

$$d = a + b * c$$

(where "*" is the multiplication operator). A programmer is free to write either the assembly code

> GET *a,*
> ADD *b,*
> MPY *c,*
> STO *d,*

which means that *c* is to be multiplied by the sum of *a* and *b* and the product stored in *d,* or he can write

> GET *b,*
> MPY *c,*
> ADD *a,*
> STO *d,*

which means that *a* is to be added to the product of *b* and *c* and the sum stored in *d*. But the higher-level language statement,

$$d = a + b * c,$$

can have only one interpretation. Which one it is to be is determined by the detailed design of the translator, hence by the designer of the language. Of course, this is not to say that the expressiveness of the higher-level language is necessarily limited. The language may, for example, permit the two different interpretations to be expressed by

$$d = (a + b) * c$$

and

$$d = a + (b * c),$$

respectively. Although this extremely simple example barely touches it, the question is not what procedures are expressible in a higher-level language—most such languages are in fact universal in Turing's sense—but what programming style the language dictates. Abraham Maslow, the psychologist, once said that to a person who has only a hammer, the whole world looks like a nail. A language is also a tool, and it also, indeed, very strongly, shapes what the world looks like to its user.

A solitary programmer writing a program to solve some problem incidental to his private research need hardly concern himself over the way his program might structure the worldview of anyone besides himself. Higher-level languages, however, are intended to be very much public languages, in two important senses: they are intended to be used by very large numbers of programmers as the languages in which they instruct their machines; and they are also intended to be used as languages in which people are to communicate to one another the procedures they have composed. Indeed, the success of a particular higher-level programming language is measured largely by the extent to which it has become dominant in the market, so to say, i.e., the extent to which it has driven competitive modes of expression into oblivion.

The recent history of the behavioral sciences has shown us how deeply the success of mathematics as used in physics has affected disciplines quite far removed from physics. Many psychologists and sociologists have for generations discussed their subject matter in terms of differential equations and statistics, for example. They may have begun by believing that the calculi they adopted were merely a convenient shorthand for describing the phenomena with which they deal. But, as they construct ever larger conceptual frameworks out of elementary components originally borrowed from foreign contexts, and as they give these frameworks names and manipulate them as elements of still more elaborate systems of thought, these frameworks cease to serve as mere modes of description and become, like Maslow's hammer, determinants of their view of the world. The design of a public language, then, is a serious task, one pregnant with consequences and thus laden with extraordinarily heavy responsibility. I shall have more to say about that later.

For the moment I wish to emphasize that a higher-level programming language, such as that represented by the preceding fragment, is, in fact, a formal language. The meanings of expressions written in it are determined, of course, by its transformation rules, and these, in turn, are embodied in the procedures that translate it into assembly and ultimately into machine language. If, therefore, one had to say what a particular higher-language program means, one would have to point ultimately to its machine-language reduction and even to the machine corresponding to it. One would have to say "it means what that machine does with that code." However, that is not how such questions are answered in reality. For the translator is itself a program and therefore transforms the computer into which it is loaded into quite another computer, namely, a computer for which that language is its machine language. There is then an equivalence, not only between formal languages and abstract games, but between these and computing machines. To put it another way, there are important universes of discourse in which distinctions between languages and their machine embodiments disappear.

Lest this be thought a "merely" philosophical point of little practical consequence, I must say immediately that not only do most of today's programmers think of the languages they use as being their machines, very nearly literally so, but many, perhaps most, have no knowledge whatever of their computer's machine language or of the content and structure of the translators that mediate between them and their computers. This observation is not made as a criticism. After all, a higher-level formal language is an abstract machine. No one is to be faulted for using a language or a machine more congenial to his purposes than is some other machine, merely because the other machine is somehow more primitive. But the observation does raise an important question: If today's programmers are largely unaware of the detailed structures of the physical machines they are using, of their languages, and of the translators that manipulate their programs, then they must also be largely ignorant of many of the arguments I have made here, particularly of those arguments concerning the universality of computers and the nature of effective procedures. How then do these programmers come to sense the power of the computer?

Their conviction that, so to say, the computer can do anything—i.e., their correct intuition that the languages available to them are, in some nontrivial sense, universal—comes largely from their impression that they can program any procedure they thoroughly understand. That impression, in turn, is based on their experience of the power of subroutines and of the reducibility of complex decision processes to hierarchies of binary (i.e., two-way branching) choices.

A subroutine is, as I have said, a program whose input-output behavior we understand without necessarily understanding how it converts the input we give it to the output it delivers. A programmer faced with a difficult subproblem can always pretend there exists a subroutine that solves that subproblem. Perhaps he will find it in a program library somewhere. But even if he doesn't, he can attend to it after he has worked out the strategy for his main problem. If, for example, his main problem is to write a program to enable a computer to play chess, he will certainly need a subroutine that will produce a chess move, given a configuration of pieces on a chess board, and an indication of whose move it is, black or white. The input to such a subroutine must be some representation of the pieces on the chess board and, say, a "0" if it is white's move or "1" if it is black's. Its output is a chess configuration in the same representation as that employed in the input. Once he has decided on suitable representations, and that may be a nontrivial task in itself, he is free to "use" a move-generating subroutine in the further development of his main program as if it existed. If he is part of a programming team, he may well describe the subroutine's desired input-output behavior to a colleague who will then write it for him. In any case, he can go on with the strategic analysis and even the programming of his main problem without having to keep lower-level details constantly in mind.

I have already said that in computers and computation larger entities are built out of smaller ones. In a way, subroutines allow us to say the opposite as well. For with their aid we can break a very large and complex task into a set of smaller tasks, and each one of these into still smaller ones, and so on. Were that not so, no one would dare to undertake such monumental tasks as computer-controlled air-traffic management or the computer simulation of a large business. Of course, the idea of solving a problem by breaking it up into smaller problems and then solving them, and so on, is far from new. But, until the advent of the computer, huge hierarchical problem-solving systems always had to depend on people to carry out the designs of the master strategists. People, unlike computers, are not altogether reliable executors of even the most explicit instructions. The computer programmer's sense of power derives largely from his conviction that his instructions will be obeyed unconditionally and that, given his ability to build arbitrarily large program structures, there is no limit to at least the size of the problems he can solve.

I am finally in a position to comment on an erroneous impression undoubtedly left by something I wrote near the beginning of Chapter II. There I wrote "machines do only what they are made to do—and that they do exactly." We have now seen how we can instruct computers to do things. If we take seriously, as we should, the statement that a program transforms one computer into another, then we have even seen how we can "make" a machine to do something. But the impression I have probably engendered is that programs must necessarily be in a form, perhaps thinly disguised, of a sequence of instructions that must be executed more or less sequentially.

I have already exhibited evidence to the contrary. The quintuples that govern the behavior of a classic Turing machine do not have to be in any specific order, nor are they obeyed sequentially. A typical such quintuple, cast in the form of a rule of a game, says;

"If you are in state u and if the symbol you are currently scanning is x, then do . . . ."

The Turing machine literally searches its tape for the rule applicable to its situation. The order in which the rules are written is totally irrelevant to the machine's ultimate input-output behavior.

There are two important points to be made in this connection. The first has to do with the idea of searching for a rule. The Turing machine searches linearly for a rule to apply. The information on its tape is, after all, in linear order. But in modern computers other searching regimens are possible. In particular, one can write programs whose function is to compute search rules. Then, just as a programmer may know a procedure for computing square roots of numbers but not know the square root of any particular number, let alone of all the numbers that may be given to his procedure, so he may know how to specify the construction of search rules but not know any of the rules his specification may produce. A computer that obeys rules it finds by carrying out such computed searching regimens does indeed do only what it has been made to do. But the "making" is at least one, possibly many more, removes from the original maker, i.e., from the programmer. I shall discuss some implications of this again.

The second point has to do with how rules weakly analogous to Turing-machine quintuples may be given to computers. Certainly rules may be entered in the form of programs of the kind I have exhibited. But let us leave computers for a moment and consider how a person would wrestle with the question, "What's the area of the dish?" First he would assume that the dish he is being asked about is the last dish mentioned. He would therefore search his memory for the last mention of a dish. If that mention was, "The area of the dish is such-and-such," he would be able to answer quickly. Otherwise he would search his memory for statements of the form, "The area of so-and-so is . . .," i.e., for some stored knowledge of how one answers questions about the area of things. Suppose he finds the statement, "The area of a circle is pi times the radius of the circle squared." He might think, "If the dish has a radius, it may be a circle and then I could apply this rule." If it turned out that the last mention of a dish was "The radius of a dish is 10," the rest would be obvious. Below is the typescript of a "conversation" with a computer conducted via an electric-typewriter computer console (the lines printed all in capitals are the computer's responses).[1]

> The area of a circle is pi times the radius of the circle squared.
> The area of a sphere is pi times the diameter of the sphere
>   squared.
> The area of a square is the side of the square squared.
> The diameter of an object is twice the radius of the object.
>
> The radius of a ball is 10.
> What is the area of the ball?
>
> THE ASSUMPTION THAT BALL IS CIRCLE IS CONSISTENT;
>   SHALL I GO ON?
> no.
> THE ASSUMPTION THAT BALL IS SPHERE AND OBJECT IS
>   CONSISTENT; SHALL I GO ON?
> yes.
> IT'S 1256.6359.

What is the diameter of the ball?
IT'S 20.

The radius of a dish is 10.
What is the area of the dish?
THE ASSUMPTION THAT DISH IS CIRCLE IS CONSISTENT;
SHALL I GO ON?
yes.
IT'S 314.15899.

The radius of a globe is 10.
A globe is a sphere. A sphere is an object.
What is the area of the globe?
IT'S 1256.6359.

The side of a plate is 10.
What is the area of the plate?
THE ASSUMPTION THAT PLATE IS SQUARE IS
CONSISTENT; SHALL I GO ON?
yes.
IT'S 100.

When I say board I mean plate.
What is 3 times the side of the board?
IT'S 30.
Please call that x.
Let $f(z)$ be z squared minus z.
Compute $f(x)$.
IT'S 870.

The program that made a computer do this is best seen, not as a set of instructions of the form "If this, do that," but as a collection of rules of evidence. It uses these rules to decide how assertions it may be given are relevant to questions it may be asked. The computer's behavior is therefore governed as much by the assertions it is fed, e.g., "A globe is a sphere," as by its instructions. The oft-repeated truism that a computer can do only what it is told to do thus turns out, like most truisms about complex matters, to be, to say the least, problematical. There are many ways to "tell" a computer something.

The idea that a person can write a program that embodies anything he "thoroughly understands" is at least equally problematical. Understanding something always means understanding it at a certain level. An actuary uses some fairly sophisticated mathematical tools whose fundamentals he almost certainly doesn't understand or care anything about. Everyone who makes change uses arithmetic, but very few people know or care much about the beautiful axiom system on which arithmetic is based. In effect, we all constantly use subroutines whose input-output behavior we believe we know, but whose details we need not and rarely do think about. To understand something sufficiently well to be able to program it for a computer does not mean to understand it to its ultimate depth. There can be no such ultimate understanding in practical affairs. Programming is rather a test of understanding. In this respect it is like writing; often when we think we understand something and attempt to write about it, our very act of composition reveals our lack of understanding even to ourselves. Our pen writes the word "because" and suddenly stops. We thought we understood the "why" of something, but discover that we don't. We begin a sentence with "obviously," and then see that what we meant to write is not obvious at all. Sometimes we connect two clauses with the word "therefore," only to then see that our chain of reasoning is defective. Programming is like that. It is, after all, writing too. But in ordinary writing we sometimes obscure our lack of understanding, our failures in logic, by unwittingly appealing to the immense flexibility of natural language and to its inherent ambiguity. The very eloquence that natural language permits sometimes illuminates our words and seems (falsely, to be sure) to illuminate our undeserving logic just as brightly. An interpreter of programming-language texts, a computer, is immune to the seductive influence of mere eloquence. And words like "obviously" are not represented in the primitive vocabularies of any computers. A computer is a merciless critic. Therefore the assertion that one understands a thing sufficiently well to be able to program it is, first of all, an assertion that one understands it in very particular terms. In any case, it can be no more than a boast that may well be falsified by experience.

The other side of the coin is the belief that one cannot program anything unless one thoroughly understands it. This misses the truth that programming is, again like any form of writing, more often than not experimental. One programs, just as one writes, not because one understands, but in order to come to understand. Programming is an act of design. To write a program is to legislate the laws for a world one first has to create in imagination. Only very rarely does any designer, be he an architect, a novelist, or whatever, have so coherent a picture of the world emergent in his imagination that he can compose its laws without criticism from that world itself. That is precisely what the computer may provide.

One must, of course, learn how to use criticism. In many cases the computer's criticism is very sharp and cannot be ignored; a program doesn't work at all, or delivers obviously wrong results. Then there is no question that something has to be redesigned. Computers are maddeningly efficient at stumbling over purely technical, i.e., linguistic, programming errors, but stumbling in a way that disguises the real locus of the trouble, e.g., just which parenthesis was misplaced. Indeed, many professional programmers believe that their craft is difficult because the languages with which they must deal have rigid syntactical rules. There is therefore a persistent cry for natural-language, e.g., English, programming systems. Programmers who hold to this belief have probably never tackled a truly difficult problem, and have therefore never felt the need for really deep criticism from the computer. It is true that in order to write one has first to master the syntactic rules of one's chosen language. But then one must also have thought through what one has to say. Literary criticism is not the business of calling attention to spelling errors and to technical violations of grammatical rules. It has to do with substantive ideas. The literary critic has to know much. A real reason that programming is very hard is that, in most instances, the computer knows nothing of those aspects of the real world that its program is intended to deal with. It is in a position analogous to, say, that of a teacher of English grammar who has been given an airplane pilot's manual and left at a plane's controls in midair. He may be a very skilled grammatical faultfinder, but as to how to fly an airplane, he knows nothing. The manual had therefore better explain everything in terms sufficiently primitive that even an earthbound schoolteacher can understand them. How much easier it would be to write a pilot's manual for an aerodynamicist who, while still unable to pilot an airplane, knew a theory of flight. It is in fact very hard to explain anything in terms of a primitive vocabulary that has nothing whatever to do with that which has to be explained. Yet that is precisely what most programs attempt to do. The difficulties that ensue are no more rooted in syntactic rigidities than is, say, the difficulty of writing a good sonnet rooted in the rigid form demanded by that class of poem. To write a good sonnet or a good program, one must know what one wants to say. And it helps enormously if one's critic shares one's relevant knowledge base.

We have seen a program that consists in part of assertions given it in natural language. These may be supplied to the program at various times. The whole set of assertions may therefore grow very large. It is, in a sense, easier to program in this style than in the more orthodox ones commonly used. And it is possible, using such programming techniques, to build a knowledge base into a computer. Such a programming system thus demonstrates that it is possible to escape from the rigidity of form commonly associated with programming.

But we mustn't make too much of this. A lazy student's list of useful physics formulas may enable him to solve many schoolbook problems in physics, but will still not give him an understanding of physics, a theory he can think about. Just so, a set of assertions of the kind we have shown may enable a computer to solve certain problems. But this says nothing of a programmer's or of a computer's mastery of a theory or of either's understanding of anything more than how to use a group of "facts" to arrive at certain conclusions. A computer's successful performance is often taken as evidence that it or its programmer understand a theory of its performance. Such an inference is unnecessary and, more often than not, is quite mistaken. The relationship between understanding and writing thus remains as problematical for computer programming as it has always been for writing in any other form.

4

SCIENCE AND THE
COMPULSIVE PROGRAMMER

There is a distinction between physically embodied machines, whose ultimate function is to transduce energy or deliver power, and abstract machines. i.e., machines that exist only as ideas. The laws which the former embody must be a subset of the laws that govern the real world. The laws that govern the behavior of abstract machines are not necessarily so constrained. One may, for example, design an abstract machine whose internal signals are propagated among its components at speeds greater than the speed of light, in clear violation of physical law. The fact that such a machine cannot actually be built does not prohibit the exploration of its behavior. It can be thought about and even simulated on a computer. (Indeed, the Education Research Center at M.I.T. has made computer-generated films that enable viewers to observe a world in which vehicles travel at physically impossible speeds.) The human imagination must be capable of transcending the limitations of physical law if it is to be able to conceive such law at all.

The computer is, of course, a physically embodied machine and, as such, cannot violate natural law. But it is not completely characterized by only its manifest interaction with the real world. Electrons flow through it, its tapes move, and its lights blink, all in strict obedience to physical law, to be sure and the courses of its internal rivers of electrons are determined by openings and closings of gates, that is, by physical events. But the game the computer plays out is regulated by systems of ideas whose range is bounded only by the limitations of the human imagination. The physically determined bounds on the electronic and mechanical events internal to the computer do not matter for that game—any more than it matters how tightly a chess player grips his bishop or how rapidly he moves it over the board.

A computer running under control of a stored program is thus detached from the real world in the same way that every abstract game is. The chess board, the 32 chessmen, and the rules of chess constitute a world entirely separate from every other world. So does a computer system together with its operating manual. The chess player who has made a bad move cannot explain his error away by pointing to some external empirical fact that he could not have known but that, had he known it, would have led him to make a better move. Neither may a programmer whose program behaves differently from what he had intended look for the fault anywhere but in the game he has himself created. He may have misread the computer system's manuals or otherwise have misunderstood his computing environment, just as a novice chess player may misread the rules for, say, castling. But no datum existing in the world outside the computer system he is using can be at all relevant to the behavior of the world he has created. A computer's failure to behave exactly as its programmer intends cannot even be attributable solely to some limitation unique to the computer. In effect, every general-purpose computer is a kind of universal machine that can in principle do what any other general-purpose computer can do. In this important sense, a specific general-purpose computer has no limitations unique to it. The computer, then, is a playing field on which one may play out any game one can imagine. One may create worlds in which there is no gravity, or in which two bodies attract each other, not by Newton's inverse-square law, but by an inverse-cube (or n^{th}-power) law, or in which time dances forward and backward in obedience to a choreography as simple or complex as one wills. One can create societies in whose economies prices rise as goods become plentiful and fall as they become scarce, and in which homosexual unions alone produce offspring. In short, one can singlehandedly write and produce plays in a theater that admits of no limitations. And, what is most important, one need know only what can be inferred directly from one's computer-system manual or constructed by one's own imagination.

By way of contrast, let us consider the act of designing a computer circuit. Earlier I displayed a circuit diagram for a one-bit adder (Figure 3.5, p. 80). One may build the corresponding adder for a machine whose basic cycle time is, say, one microsecond, in other words, for a machine that experiences a million quiescent-active time-interval pairs every second. Suppose such an adder is built and is found to function properly. Say the adder's designers later install it in a machine that operates at ten times the rate of the original machine, that, so to speak, turns itself on and off ten million times per second, and suppose that the adder doesn't work in its new environment. What can be the trouble? More importantly, what sources of knowledge may have to be tapped in order to correctly diagnose the disorder? Clearly the rules of the game as stated by the abstract equations governing the behavior of AND, OR, and NOT gates and by the adder's circuit diagram are not sufficient. Nothing has changed with respect to the adder's abstract design. By hypothesis, all that has changed is the rate at which the adder is being exercised. The new rate must therefore be responsible for the adder's malfunctioning. In order to discover how to repair the adder or, indeed, to know whether it is physically possible to operate any device at such a high rate at all, one must apply some relevant knowledge of the physical world.

This example, though simplified, is not fanciful, and it is worth pursuing a little further. Nearly everyone knows that violin strings (and other objects as well) have natural (or resonant) frequencies of vibration. The A string, for example, vibrates naturally approximately 435 times per second. If two A strings are near one another, and one of them is made to vibrate at its natural frequency, say, by plucking it, then the other will also be set into vibration (at, of course, the same frequency). If, however, the first string is forced to vibrate at some frequency other than its resonant one, then, even though both are A strings, the second will not vibrate. In effect, strings are both transmitting and receiving antennas at their natural acoustic frequencies. With this in mind, suppose two children living across the street from one another set up a signaling system consisting of two parallel strings bridging the street and connecting their two houses. To signal the arrival of father, say, one of the strings is made to vibrate, as is the other string to signal the arrival of mother. If, however, one string is made to vibrate at its natural frequency, then the other string will be induced to vibrate as well. The system will therefore fail to function as intended; it will be incapable of announcing the arrival of only one parent.

Electric circuits are somewhat like acoustic systems in this respect, although the frequencies associated with them are very much higher than those of their acoustic counterparts. In particular, circuits have natural (or resonant) frequencies. Indeed, many may be tuned to resonate at quite specifically chosen frequencies, in order to act, for example, as antennas for broadcasting stations. Home radio receivers have similarly tunable circuits that act as receiving antennas. The trouble with our adder may be that, although none of its circuits resonated so as to broadcast and receive signals from one another when they were operated at a frequency of a million pulses per second, they did act as broadcasting and receiving antennas at the higher operating frequency. They consequently confused the information they were designed to manipulate.

Properly trained electrical engineers, of course, command the theory from which such phenomena can be deduced. Others may identify the trouble on the basis of earlier experiences. But the engineer who has neither the theory nor the experience could never discover the cause of the trouble by deductive logic alone. He could repair his circuit only by fortuitous tinkering, or by receiving outside help, or by, in effect, repeating the creative scientific work of a man like Heinrich Hertz, the discoverer of the electromagnetic radiation phenomenon to which we are here alluding. But we label Hertz's work creative precisely because it involves observations of phenomena not mentioned in any existing manuals and because it infers general laws from the particulars observed. That is quite the opposite of deduction.

An engineer is inextricably impacted in the material world. His creativity is confined by its laws; he may, finally, do only what may lawfully be done. But he is doomed to exercise his trade in a Kafkaesque castle from which there is, even in principle, no escape. For he cannot know the whole plan that determines what rooms there are in the world, what doors exist between them, and how the doors may be opened. When a device an engineer has designed doesn't work, he therefore cannot always know, or tell by his own reasoning alone, whether he is in an antechamber to success where only his blunders keep him from opening its doors, or whether he has wandered into a closet from which there is no exit. Then he must appeal to others, his teachers, his colleagues, his books, to tell him, or at least to hint at, a formula that will compel the insouciant attendant (nature) to lead him out and on.

The computer programmer, however, is a creator of universes for which he alone is the lawgiver. So, of course, is the designer of any game. But universes of virtually unlimited complexity can be created in the form of computer programs. Moreover, and this is a crucial point, systems so formulated and elaborated *act out* their programmed scripts. They compliantly obey their laws and vividly exhibit their obedient behavior. No playwright, no stage director, no emperor, however powerful, has ever exercised such absolute authority to arrange a stage or a field of battle and to command such unswervingly dutiful actors or troops.

One would have to be astonished if Lord Acton's observation that power corrupts were not to apply in an environment in which omnipotence is so easily achievable. It does apply. And the corruption evoked by the computer programmer's omnipotence manifests itself in a form that is instructive in a domain far larger that the immediate environment of the computer. To understand it, we will have to take a look at a mental disorder that, while actually very old, appears to have been transformed by the computer into a new genus: the compulsion to program.

Wherever computer centers have become established, that is to say, in countless places in the United States, as well as in virtually all other industrial regions of the world, bright young men of disheveled appearance, often with sunken glowing eyes, can be seen sitting at computer consoles, their arms tensed and waiting to fire their fingers, already poised to strike, at the buttons and keys on which their attention seems to be as riveted as a gambler's on the rolling dice. When not so transfixed, they often sit at tables strewn with computer printouts over which they pore like possessed students of a cabalistic text. They work until they nearly drop, twenty, thirty hours at a time. Their food, if they arrange it, is brought to them: coffee, Cokes, sandwiches. If possible, they sleep on cots near the computer. But only for a few hours—then back to the console or the printouts. Their rumpled clothes, their unwashed and unshaven faces, and their uncombed hair all testify that they are oblivious to their bodies and to the world in which they move. They exist, at least when so engaged, only through and for the computers. These are computer bums, compulsive programmers. They are an international phenomenon.

How may the compulsive programmer be distinguished from a merely dedicated, hard-working professional programmer? First, by the fact that the ordinary professional programmer addresses himself to the problem to be solved, whereas the compulsive programmer sees the problem mainly as an opportunity to interact with the computer. The ordinary computer programmer will usually discuss both his substantive and his technical programming problem with others. He will generally do lengthy preparatory work, such as writing and flow diagramming, before beginning work with the computer itself. His sessions with the computer may be comparatively short. He may even let others do the actual console work. He develops his program slowly and systematically. When something doesn't work, he may spend considerable time away from the computer, framing careful hypotheses to account for the malfunction and designing crucial experiments to test them. Again, he may leave the actual running of the computer to others. He is able, while waiting for results from the computer, to attend to other aspects of his work, such as documenting what he has already done. When he has finally composed the program he set out to produce, he is able to complete a sensible description of it and to turn his attention to other things. The professional regards programming as a means toward an end, not as an end in itself. His satisfaction comes from having solved a substantive problem, not from having bent a computer to his will.

The compulsive programmer is usually a superb technician, moreover, one who knows every detail of the computer he works on, its peripheral equipment, the computer's operating system, etc. He is often tolerated around computer centers because of his knowledge of the system and because he can write small subsystem programs quickly, that is, in one or two sessions of, say, twenty hours each. After a time, the center may in fact be using a number of his programs. But because the compulsive programmer can hardly be motivated to do anything but program, he will almost never document his programs once he stops working on them. A center may therefore come to depend on him to teach the use of, and to maintain, the programs that he wrote and whose structure only he, if anyone, understands. His position is rather like that of a bank employee who doesn't do much for the bank, but who is kept on because only he knows the combination to the safe. His main interest is, in any case, not in small programs, but in very large, very ambitious systems of programs. Usually the systems he undertakes to build, and on which he works feverishly for perhaps a month or two or three, have very grandiose but extremely imprecisely stated goals. Some examples of these ambitions are: new computer languages to facilitate man-machine communication; a general system that can be taught to play any board game; a system to make it easier for computer experts to write super-systems (this last is a favorite). It is characteristic of many such projects that the programmer can long continue in the conviction that they demand knowledge about nothing but computers, programming, etc. And that knowledge he, of course, commands in abundance. Indeed, the point at which such work is often abandoned is precisely when it ceases to be purely incestuous, i.e., when programming would have to be interrupted in order that knowledge from outside the computer world may be acquired.

Unlike the professional, the compulsive programmer cannot attend to other tasks, not even to tasks closely related to his program, during periods when he is not actually operating the computer. He can barely tolerate being away from the machine. But when he is nevertheless forced by circumstances to be separated from it, at least he has his computer printouts with him. He studies them, he talks about them to anyone who will listen—though, of course, no one else can understand them. Indeed, while in the grip of his compulsion, he can talk of nothing but his program. But the only time he is, so to say, happy is when he is at the computer console. Then he will not converse with anyone but the computer. We will soon see what they converse about.

The compulsive programmer spends all the time he can working on one of his big projects. "Working" is not the word he uses; he calls what he does "hacking." To hack is, according to the dictionary, "to cut irregularly, without skill or definite purpose; to mangle by or as if by repeated strokes of a cutting instrument." I have already said that the compulsive programmer, or hacker as he calls himself, is usually a superb technician. It seems therefore that he is not "without skill" as the definition would have it. But the definition fits in the deeper sense that the hacker is "without definite purpose": he cannot set before himself a clearly defined long-term goal and a plan for achieving it, for he has only technique, not knowledge. He has nothing he can analyze or synthesize; in short, he has nothing to form theories about. His skill is therefore aimless, even disembodied. It is simply not connected with anything other than the instrument on which it may be exercised. His skill is like that of a monastic copyist who, though illiterate, is a first-rate calligrapher. His grandiose projects must therefore necessarily have the quality of illusions, indeed, of illusions of grandeur. He will construct the one grand system in which all other experts will soon write their systems.

(It has to be said that not all hackers are pathologically compulsive programmers. Indeed, were it not for the often, in its own terms, highly creative labor of people who proudly claim the title "hacker," few of today's sophisticated computer time-sharing systems, computer language translators, computer graphics systems, etc., would exist.)

Programming systems can, of course, be built without plan and without knowledge, let alone understanding, of the deep structural issues involved, just as houses, cities, systems of dams, and national economic policies can be similarly hacked together. As a system so constructed begins to get large, however, it also becomes increasingly unstable. When one of its subfunctions fails in an unanticipated way, it may be patched until the manifest trouble disappears. But since there is no general theory of the whole system, the system itself can be only a more or less chaotic aggregate of subsystems whose influence on one another's behavior is discoverable only piecemeal and by experiment. The hacker spends part of his time at the console piling new subsystems onto the structure he has already built—he calls them "new features"—and the rest of his time in attempts to account for the way in which substructures already in place misbehave. That is what he and the computer converse about.

The psychological situation the compulsive programmer finds himself in while so engaged is strongly determined by two apparently opposing facts: first, he knows that he can make the computer do anything he wants it to do; and second, the computer constantly displays undeniable evidence of his failures to him. It reproaches him. There is no escaping this bind. The engineer can resign himself to the truth that there are some things he doesn't know. But the programmer moves in a world entirely of his own making. The computer challenges his power, not his knowledge.

Indeed, the compulsive programmer's excitement rises to its highest, most feverish pitch when he is on the trail of a most recalcitrant error, when everything ought to work but the computer nevertheless reproaches him by misbehaving in a number of mysterious, apparently unrelated ways. It is then that the system the programmer has himself created gives every evidence of having taken on a life of its own and, certainly, of having slipped from his control. This too is the point at which the idea that the computer can be "made to do anything" becomes most relevant and most soundly based in reality. For, under such circumstances, the misbehaving artifact is, in fact, the programmer's own creation. Its very misbehavior can, as we have already said, be the consequence only of what the programmer himself has done. And what he has done he can presumably come to understand, to undo, and to redo to better serve his purpose. Accordingly his mood and his activity become frenzied when he believes he has finally discovered the source of the trouble. Should his time at the console be nearly up at that moment, he will take enormous risks with his program, making substantial changes, one after another, in minutes or even seconds without so much as recording what he is doing, always pleading for just another minute. He can, under such circumstances, rapidly and virtually irretrievably destroy weeks and weeks of his own work. Should he, however, find a deeply embedded error, one that actually does account for much of the program's misbehavior, his joy is unbounded. It *is* a thrill to see a hitherto moribund program suddenly come back to life; there is no other way to say it. When some deep error has been found and repaired, many different portions of the program, which until then had given nothing but incomprehensible outputs, suddenly behave smoothly and deliver precisely the intended results. There is reason for the diagnostician to be pleased and, if the error was really deep inside the system, even proud.

But the compulsive programmer's pride and elation are very brief. His success consists of his having shown the computer who its master is. And having demonstrated that he can make it do this much, he immediately sets out to make it do even more. Thus the entire cycle begins again. He begins to "improve" his system, say, by making it run faster, or by adding "new features" to it, or by improving the ease with which data can be entered into it and gotten out of it. The act of modifying the then-existing program invariably causes some of its substructures to collapse; they constitute, after all, an amorphous collection of processes whose interactions with one another are virtually fortuitous. His apparently devoted efforts to improve and promote his own creation are really an assault on it, an assault whose only consequence can be to renew his struggle with the computer. Should he be prevented from so sabotaging his own work, say, by administrative decision, he will become visibly depressed, begin to sulk, display no interest in anything around him, etc. Only a new opportunity to compute can restore his spirit.

It must be emphasized that the portrait I have drawn is instantly recognizable at computing installations all over the world. It represents a psychopathology that is far less ambiguous than, say, the milder forms of schizophrenia or paranoia. At the same time, it represents an extremely developed form of a disorder that afflicts much of our society.

How are we to understand this compulsion? We must first recognize that it *is* a compulsion. Normally, wishes for satisfaction lead to behaviors that have a texture of discrimination and spontaneity. The fulfillment of such wishes leads to pleasure. The compulsive programmer is driven; there is little spontaneity in how he behaves; and he finds no pleasure in the fulfillment of his nominal wishes. He seeks reassurance from the computer, not pleasure. The closest parallel we can find to this sort of psychopathology is in the relentless, pleasureless drive for reassurance that characterizes the life of the compulsive gambler.

The compulsive gambler is also to be sharply distinguished from the professional gambler. The latter is, in an important sense, not a gambler at all. (We may leave aside the cheater and the professional confidence man, for certainly neither of them are gamblers either.) The so-called professional gambler is really an applied statistician, and perhaps an applied psychologist as well. His income depends in almost no way on luck. He knows applied probability theory, and he uses it to calculate odds and then to play those odds in such combinations and aggregates that he can predict his income during a period of, say, a year with almost mathematical certainty. That is not gambling. Then there are people who gamble but who are neither professional nor compulsive gamblers. To the compulsive gambler, gambling, the game, is everything. Even winning is less important than playing. He is, so to say, happy only when he is at the gambling table.

Anyone who has ever worked in a computer center or a gambling casino that closes its doors at night will recognize the scene described by Dostoevski, himself a passionate gambler, in *The Gambler:*

> "By eleven o'clock, there remain at the roulette table only those desperate players, the real gamblers, for whom there exists but the roulette table, . . . who know nothing of what is going on around them and take no interest in any matters outside the roulette saloon, but only play and play from morning till night, and would gladly play all round the clock if it were permitted. These people are always annoyed when midnight comes, and they must go home, because the roulette bank is closed. And when the chief croupier, about 12 o'clock, just before the close calls out, 'The last three turns, gentlemen!' these men are ready to stake all they have in their pockets on those last three turns, and it is certain that it is just then that these people lose most."[1]

Dostoevski might as well have been describing a computer room.

The medical literature on compulsive gambling concerns itself mainly with the psychogenesis of that compulsion, and then almost entirely from a psychoanalytic perspective. But I need not recapitulate the psychoanalytic argument for my purposes here. It is enough to say here that psychoanalysts, beginning with Freud, saw megalomania and fantasies of omnipotence as principal ingredients in the psychic life of the compulsive gambler. We do not have to accept, or reject, psychoanalytic accounts of the origins of such fantasies—e.g., that they are rooted in unresolved Oedipal conflicts leading to wishes to overpower the father that in turn lead to unconscious motivations to fail—in order to join the psychoanalysts and novelists like Dostoevski in seeing the central role that megalomaniac fantasies of omnipotence play in compulsive gambling.

The gambler, according to the psychoanalyst Edmund Bergler, has three principal convictions: first, he is subjectively certain that he will win; second, he has an unbounded faith in his own cleverness; third, he knows that life itself is nothing but a gamble.[2]

What grounds can there possibly be for knowing that one will win a game of pure chance? To know that the roll of a pair of dice or the turn of a card is a purely chance event is to know that nothing one does can possibly affect the outcome. There precisely is the rub! The compulsive gambler believes himself to be in control of a magical world to which only few men are given entrance. "He believes," writes Bergler, "Fate has singled him out . . . and communicates with him by means of small signs indicating approval and reproach."[3] The gambler is the scientist of this magical world. He is the interpreter of the signs that Fate communicates to him, just as the scientist in the real world is an interpreter of the signs that nature communicates to everyone who cares to become sensitive to them. And like the natural scientist, the compulsive gambler always has a tentative hypothesis that accounts for almost all the signs he has so far observed, that, in other words, constitutes a very nearly complete picture of those aspects of the universe which interest him. The test of the adequacy of both the scientist's and the magician's view of the world is its power to predict and, under suitably arranged conditions, to control. Hence, according to Bergler, the compulsive gambler sees himself as "not the victim, but the executive arm of unpredictable chance."[4]

What an outsider regards as the gambler's superstitions are in fact manifestations of the gambler's hypothetical reconstruction of the world Fate has bit by bit revealed to him. Experience has taught him, say, that in order to win he must touch a hunchback on the day of play, carry a rabbit's foot in his left pocket, not sit at the gaming table with his legs crossed, and so on. This sort of knowledge is to him what, say, the knowledge of the mathematics of airflow over wings may be to an aircraft designer.

Because the gambler's superstitions are effectively irrelevant to the motions of dice, the orderings of cards, and so on, his hypotheses are very often empirically falsified. Each falsifying experience, however, contains certain elements that can be integrated into the main lines of his hypothetical framework and so save its over-all structure. Losing, therefore, doesn't mean that carrying a rabbit's foot, for example, is wrong or irrelevant, but only that some crucial ingredient for success has so far been overlooked. Perhaps the last time the gambler did win, a blond young lady stood behind his chair. Ah! So that's it: Touch a hunchback, carry a rabbit's foot, don't cross legs, *and* have a blond young lady stand behind the chair. When that doesn't work, he calculates that that particular combination works only on Thursdays, and so on and on and on. Bits and pieces of explanation are added on, some are removed, and the entire structure becomes more and more complicated. Eventually, the gambler really does command a conceptual framework that rivals a body of scientific knowledge, at least in its complexity and intricacy. He is an expert in a richly complicated world open only to the few initiates who have, through their own hard work and risktaking, learned its mysterious lore and language.

The magical world inhabited by the compulsive gambler is no different in principle from that in which others, equally driven by grandiose fantasies, attempt to realize their dreams of power. Astrology, for example, has constructed an enormously complex conceptual framework, a system of theories and hypotheses which allegedly permit the cognizant to control events. To know, for example, that the conjunction of certain planets on a particular date bodes ill for a particular venture, but that some other conjunction on some other date bodes well for it, and then to undertake that venture on the favored date, is to attempt to control events.

But the hypotheses of astrology too are routinely falsified by events. How then does astrology, and how do other such magical systems, remain at all a force in the minds of men? Exactly as do the hypotheses of the compulsive gambler. First, any contradiction between experience and one magical notion is explained by reference to other magical notions. Thus the entire structure of the magical system of beliefs is supported by its very circularity. This way of protecting the system against assaults by reality is especially effective if objections are always met one at a time, for then the very demonstration that an apparently anomalous fact can be incorporated into the system serves to validate the system. The gambler may, for example, appeal to the fact that he didn't tie his shoelace, as he knew he should have, to account for his "bad luck" on a particular day. That sort of explanation is formally equivalent to the compulsive programmer's assumption that his program's misbehavior is caused by a merely technical programming error.

A second way in which conceptual frameworks of gamblers and of programmers are protected is by cyclical elaboration. The gambler who suddenly realizes that certain of his tricks work only on Thursdays simply incorporates this new "insight" into his already existing framework of superstitions, thus, in effect, adding an epicycle to its structure. The programmer is free to convert every new embarrassment into a special case to be handled by a specially constructed, ad hoc subprogram and to be thus incorporated into his over-all system. Such unbounded epicyclic elaborations of their systems provide both programmers and gamblers with an inexhaustible reserve of subsidiary explanations for even the gravest difficulties.

Finally, the conceptual stability of a magical or programming system may be protected by denying, to use the words of Michael Polanyi,

> "to any rival conception the ground in which it might take root. Experiences which support [the rival conception] could be adduced only one by one. But a new conception . . . which could take the place of [the one held] could be established only by a whole series of relevant instances, and such evidence cannot accumulate in the minds of [gamblers or programmers] if each [bit of evidence] is disregarded in its turn for lack of the concept which would lend significance to it."[5]

The gambler constantly defies the laws of probability; he refuses to recognize their operational significance. He therefore cannot permit them to become a kernel of a realistic insight. A particular program may be foundering on deep structural, mathematical, or linguistic difficulties about which relevant theories exist. But the compulsive programmer meets most manifestations of trouble with still more programming tricks, and thus, like the gambler, refuses to permit them to nucleate relevant theories in his mind. Compulsive programmers are notorious for not reading the literature of the substantive fields in which they are nominally working.

These three mechanisms, called by Polanyi circularity, self-expansion, and suppressed nucleation, constitute the main defensive armamentarium of the true adherent of magical systems of thought, and particularly of the compulsive programmer. Psychiatric literature informs us that this pathology deeply involves fantasies of omnipotence. The conviction that one is all-powerful, however, cannot rest; it must constantly be verified by tests. The test of power is control. The test of absolute power is certain and absolute control. When dealing with the compulsive programmer, we are therefore also dealing with his need to control and his need for certainty.

The passion for certainty is, of course, also one of the great cornerstones of science, philosophy, and religion. And the quest for control is inherent in all technology. Indeed, the reason we are so interested in the compulsive programmer is that we see no discontinuity between his pathological motives and behavior and those of the modern scientist and technologist generally. The compulsive programmer is merely the proverbial mad scientist who has been given a theater, the computer, in which he can, and does, play out his fantasies.

Let us reconsider Bergler's three observations about gamblers. First, the gambler is subjectively certain that he will win. So is the compulsive programmer—only he, having created his own world on a universal machine, has some foundation in reality for his certainty. Scientists, with some exceptions, share the same faith: what science has not done, it has not *yet* done; the questions science has not answered, it has not *yet* answered. Second, the gambler has an unbounded faith in his own cleverness. Well?! Third, the gambler knows that life itself is nothing but a gamble. Similarly, the compulsive programmer is convinced that life is nothing but a program running on an enormous computer, and that therefore every aspect of life can ultimately be explained in programming terms. Many scientists (again there are notable exceptions) also believe that every aspect of life and nature can finally be explained in exclusively scientific terms. Indeed, as Polanyi correctly points out, the stability of scientific beliefs is defended by the same devices that protect magical belief systems:

> "Any contradiction between a particular scientific notion and the facts of experience will be explained by other scientific notions; there is a ready reserve of possible scientific hypotheses available to explain any conceivable event. . . . *within science itself*, the stability of theories against experience is maintained by epicyclical reserves which suppress alternative conceptions in the germ."[6]

Hence we can make out a continuum. At one of its extremes stand scientists and technologists who much resemble the compulsive programmer. At the other extreme are those scientists, humanists, philosophers, artists, and religionists who seek understanding as whole persons and from all possible perspectives. The affairs of the world appear to be in the hands of technicians whose psychic constitutions approximate those of the former to a dangerous degree. Meanwhile the voices that speak the wisdom of the latter seem to be growing ever fainter.

There is a well-known joke that may help clarify the point. One dark night a policeman comes upon a drunk. The man is on his knees, obviously searching for something under a lamppost. He tells the officer that he is looking for his keys, which he says he lost "over there," pointing out into the darkness. The policeman asks him "Why, if you lost the keys over there, are you looking for them under the streetlight?" The drunk answers, "Because the light is so much better here." That is the way science proceeds too. It is important to recognize this fact, irrelevant and useless to blame science for it. Indeed, what is sought can be found only where there is illumination. Sometimes one even finds a new source of light in the circle within which one is searching. Two things matter: the size of the circle of light that is the universe of one's inquiry, and the spirit of one's inquiry. The latter must include an acute awareness that there is an outer darkness, and that there are sources of illumination of which one as yet knows very little.

Science can proceed only by simplifying reality. The first step in its process of simplification is abstraction. And abstraction means leaving out of account all those empirical data which do not fit the particular conceptual framework within which science at the moment happens to be working, which, in other words, are not illuminated by the light of the particular lamp under which science happens to be looking for keys. Aldous Huxley remarked on this matter with considerable clarity:

> "Pragmatically [scientists] are justified in acting in this odd and extremely arbitrary way; for by concentrating exclusively on the measurable aspects of such elements of experience as can be explained in terms of a causal system they have been able to achieve

a great and ever increasing control over the energies of nature. But power is not the same thing as insight and, as a representation of reality, the scientific picture of the world is inadequate for the simple reason that science does not even profess to deal with experience as a whole, but only with certain aspects of it in certain contexts. All this is quite clearly understood by the more philosophically minded men of science. But unfortunately some scientists, many technicians, and most consumers of gadgets have lacked the time and the inclination to examine the philosophical foundations and background of the sciences. Consequently they tend to accept the world picture implicit in the theories of science as a complete and exhaustive account of reality; they tend to regard those aspects of experience which scientists leave out of account, because they are incompetent to deal with them, as being somehow less real than the aspects which science has arbitrarily chosen to abstract from out of the infinitely rich totality of given facts."[7]

One of the most explicit statements of the way in which science deliberately and consciously plans to distort reality, and then goes on to accept that distortion as a "complete and exhaustive" account, is that of the computer scientist Herbert A. Simon, concerning his own fundamental theoretical orientation:

"An ant, viewed as a behaving system, is quite simple. The apparent complexity of its behavior over time is largely a reflection of the complexity of the environment in which it finds itself. . . . the truth or falsity of [this] hypothesis should be independent of whether ants, viewed more microscopically, are simple or complex systems. At the level of cells or molecules, ants are demonstrably complex; but these microscopic details of the inner environment may be largely irrelevant to the ant's behavior in relation to the outer environment. That is why an automaton, though completely different at the microscopic level, might nevertheless simulate the ant's gross behavior. . . .

"I should like to explore this hypothesis, but with the word 'man' substituted for 'ant.'

"A man, viewed as a behaving system, is quite simple. The apparent complexity of his behavior over time is largely a reflection of the complexity of the environment in which he finds himself. . . . I myself believe that the hypothesis holds even for the whole man."[8]

With a single stroke of the pen, by simply substituting "man" for "ant," the presumed irrelevancy of the microscopic details of the ant's inner environment to its behavior has been elevated to the irrelevancy of the whole man's inner environment to his behavior! Writing 23 years before Simon, but as if Simon's words were ringing in his ears, Huxley states;

> Because of the prestige of science as a source of power, and because of the general neglect of philosophy, the popular Weltanschauung of our times contains a large element of what may be called 'nothing-but' thinking. Human beings, it is more or less tacitly assumed, are nothing but bodies, animals, even machines. . . . values are nothing but illusions that have somehow got themselves mixed up with our experience of the world; mental happenings are nothing but epiphenomena. . . . spirituality is nothing but . . . and so on."[9]

Except, of course, that here we are not dealing with the "popular" Weltanschauung, but with that of one of the most prestigious of American scientists. Nor is Simon's assumption of what is irrelevant to the whole man's behavior "more or less tacit"; to the contrary, he has, to his credit, made it quite explicit.

Simon also provides us with an exceptionally clear and explicit description of how, and how thoroughly, the scientist prevents himself from crossing the boundary between the circle of light cast by his own presuppositions and the darkness beyond. In discussing how he went about testing the theses that underly his hypothesis, i.e., that man is quite simple, etc., he writes;

> "I have surveyed some of the evidence from a range of human performances, particularly those that have been studied in the psychological laboratory.
>
> "The behavior of human subjects in solving cryptarithmetic problems, in attaining concepts, in memorizing, in holding information in short-term memory, in processing visual stimuli, and in performing tasks that use natural languages provides strong support for these theses. . . . generalizations about human thinking . . . are emerging from the experimental evidence. They are simple things, just as our hypothesis led us to expect. Moreover,

though the picture will continue to be enlarged and clarified, we should not expect it to become essentially more complex. Only human pride argues that the apparent intricacies of our path stem from a quite different source than the intricacy of the ant's path."[10]

The hypothesis to be tested here is, in part, that the inner environment of the whole man is irrelevant to his behavior. One might suppose that, in order to test it, evidence that might be able to falsify it would be sought. One might, for example, study man's behavior in the face of grief or of a profound religious experience. But these examples do not easily lend themselves to the methods for the study of human subjects developed in psychological laboratories. Nor are they likely to lead to the simple things an experimenter's hypotheses lead him to expect. They lie in the darkness in which the theorist, in fact, has lost his keys; but the light is so much better under the lamppost he himself has erected.

There is thus no chance whatever that Simon's hypothesis will be falsified in his or his colleagues' minds. The circle of light that determines and delimits his range of vision simply does not illuminate any areas in which questions of, say, values or subjectivity can possibly arise. Questions of that kind, being, as they must be, entirely outside his universe of discourse, can therefore not lead him out of his conceptual framework, which, like all other magical explanatory systems, has a ready reserve of possible hypotheses available to explain any conceivable event.

Almost the entire enterprise that is modern science and technology is afflicted with the drunkard's search syndrome and with the myopic vision which is its direct result. But, as Huxley also pointed out, this myopia cannot sustain itself without being nourished by experiences of success. Science and technology are sustained by their translations into power and control. To the extent that computers and computation may be counted as part of science and technology, they feed at the same table. The extreme phenomenon of the compulsive programmer teaches us that computers have the power to sustain megalomaniac fantasies. But that power of the computer is merely an extreme version of a power that is inherent in all self-validating systems of thought. Perhaps we are beginning to understand that the abstract systems—the games computer people can generate in their infinite freedom from the constraints that delimit the dreams of workers in the real world—may fail catastrophically when their rules are applied in earnest. We must also learn that the same danger is inherent in other magical systems that are equally detached from authentic human experience, and particularly in those sciences that insist they can capture the *whole man* in their abstract skeletal frameworks.

5

THEORIES AND MODELS

Suppose a team of explorers from a highly technological society just like ours, but one that knew nothing about computers, were to come upon a functioning computer. They find that they cannot break into it, can gain no access to its, so to say, electro-neurophysiological apparatus. They do notice, however, that whenever they type something on its console typewriter, the computer's lights flash in a complex but apparently orderly way, its magnetic tapes sometimes spin, and the typewriter types a message that appears to be a response to what they have typed. After a time, they discover that they can dismount the computer's magnetic tapes and cause their contents to be printed on another device, a high-speed printer, which they also find on the site of their exploration. These contents prove to be readable, at least in the sense that they are represented in the explorer's own alphabet.

Since this machine—and the explorers do recognize it as a machine—is obviously a behaving instrument, the explorers naturally wish to discover the laws of its behavior. How could they go about reaching the understanding they desire? Indeed, what can it mean to understand the machine's laws of behavior?

We, of course, can put ourselves in the position of a highly privileged observer, somewhat like that of a chemistry instructor who knows very well what compound he gave his students to analyze. We know that the machine the explorers found is a computer, moreover, a computer of precisely such and such a type and one containing a particular program we also know in detail. We can therefore tell the precise degree, so to say, of understanding the explorers will have achieved at any given stage of their research. If, for example, they were to produce a computer of their own which, as seen from our privileged perspective, appears to be an exact copy of the computer they found and which even contains the same program as the original, then we would have to say that they understood the original computer at least as well as did its designers.

However, lesser achievements would also deserve to be called understanding of a very high degree. Suppose, for example, that the explorers managed to build a computer whose internal structure and whose internal components are entirely different, but whose input-output behavior is indistinguishable from that of the original; in other words, no test short of breaking open either computer can determine which of the two computers, the one the explorers found or the one they built, generated what response to what input. It may be that the internal components of the found machine are made of bailing wire, chewing gum, and adhesive tape, whereas those of the explorers' functional copy are all electronic; that doesn't matter as long as, for any reason, the original machine may not be opened for detailed internal inspection. (Actually, of course, such an achievement is impossible in principle. It may be, for example, that the original computer was so constructed that it prefaces its first console typewriter output with an exclamation mark on and only on the seventeenth Thursday of every leapyear. Even if that were discovered, the explorers could never be sure that there are not other oddities which, though systematic, have not yet been discovered.

Chapter 5

We, as privileged observers, would, of course, know about such things.)

A still lesser degree of understanding could be claimed if the explorers succeeded in building some sort of digital computer, say, a simple universal Turing machine of the type we discussed in Chapter II, and then explained the machine they found in terms of Turing-machine principles. They could then account for the found machine's extraordinary versatility and even for the fact, say, that it takes it longer to compute the inverse of a large matrix than that of a smaller one. They might, on the other hand, be utterly unable to explain why it takes the found computer longer to execute algorithms given to it in one programming language than it does to execute those same algorithms written in another programming language. We, given our omniscience about computers, know that the difference in execution speeds is due to the fact that the computer translates programs in the first language a line at a time into its machine language, and then obeys the so-generated machine-language instructions, whereas it first translates the whole program written in the second language into its machine language and only then executes the entire set of so-generated machine-language instructions. The latter process is almost always much less time-consuming than the former. Presumably the explorers would eventually develop some explanation for this and other puzzling phenomena. They might, for example, conjecture that some programming languages are more familiar to the machine than some others, and might even develop some taxonomy of programming languages based on the machine's experimentally discovered familiarity with them. The concept of familiarity, as well as the taxonomy of languages for which it serves as an organizing principle, would then become part of their computer science. It is, of course, a concept weak in explanatory power, even a misleading one. But then it is much easier for us privileged observers to know this than it is for the explorers, faced as they are with the task of having to explain phenomena of unbounded complexity.

Let us press this fantasy one more step: Suppose the explorers found not just one computer, but many computers of many diverse types, all of them, however, so-called single-address machines.

Recall that a single-address machine is one whose built-in instructions have the form "operation code; address of datum to be operated on" (see p. 86). With luck, and if the explorers were clever, they would discover what they would undoubtedly call a "language universal" with respect to the grammatical structure of the machine languages they have encountered. And to explain it as something other than a mere accident (which would, of course, be no explanation at all), they would have to conclude that this universal feature of all the languages they have observed must be due to some correspondingly universal feature, some innate property, of the machines themselves. And they would, as we privileged observers know, of course, be correct; the fact that a computer's machine language has the single-address format is a direct consequence of its design. Indeed, if we assume that the machines the explorers discovered are ordinary computers and not robots—that is, that they don't have perceptors, like television eyes, and effectors, like mechanical arms and hands—then all the discoveries the explorers make and all the theories they develop must be based solely on observations of the, so to say, verbal behavior of the machines. Apart from such minor, though possibly not unhelpful, phenomena as the flashing of the computers' lights and the occasional motions of their tape reels, the only evidence of their structures that the computers provide is, after all, linguistic. They accept strings of linguistic inputs in the form of the texts typed on their console typewriters, and they respond with linguistic outputs written on the same instrument or onto magnetic tapes.

In Chapters II and III we were very much concerned with legal moves in abstract games and grammatical constructions in abstract languages. My aim there was to build up the idea of a computer on the basis of such concepts. In the fantasy we are currently entertaining, we are, in effect, looking at the other side of the same coin. We now see that, if we strive to explain computers when bounded by the restriction that we may not break the computer open, then all explanations must be derived from linguistic bases.

The position of a human being observing another human being is not so very different from that of the explorers who wish to understand the computers they have encountered. We too have extremely limited access to the neurophysiological material that appears to determine how we think. Besides, it wouldn't advance our current understanding of thinking very much even if we could subject the living brain to the kind of analysis to which we actually can subject a running computer, that is, by tracing connections, electrical pulses, and so on. Our ignorance of brain function is currently so very nearly total that we could not even begin to frame appropriate research strategies. We would stand before the open brain, fancy instruments in hand, roughly as an unschooled laborer might stand before the exposed wiring of a computer: awed perhaps, but surely helpless. A microanalysis of brain functions is, moreover, no more useful for understanding anything about thinking than a corresponding analysis of the pulses flowing through a computer would be for understanding what program the computer is running. Such analyses would simply be at the wrong conceptual level. They might help to decide crucial experiments, but only after such experiments had been designed on the basis of much higher-level (for example, linguistic) theories.

Because, in fact, scientists do suffer from the same sort of handicaps as we imposed on our mythical explorers, and cannot communicate with an omniscient observer who could, if he but would, reveal all secrets, it is not surprising that at least some scientists seek understanding of the way humans work in somewhat the same way as our explorers might have sought to understand the computers they found, that is, by designing computers whose imput-output behavior resembles that of humans as closely as possible.

The work of linguists—for example, that of Noam Chomsky—should be mentioned here, even though it does not involve the use of computers. A simpleminded and grossly misleading view of the task that Chomsky's school has set itself is that it is to systematically record the grammatical rules of as many natural languages (e.g., English) as possible. If that were the only, or even the principal, aim of Chomsky's school, we would expect it to publish a series of books, all independent of one another, entitled *"The Grammar of X,"* where *X* stands for one of the various known human languages. In fact, Chomsky's most profoundly significant working hypothesis is that man's genetic endowment gives him a set of highly specialized abilities and imposes on him a corresponding set of restrictions which, taken together, determine the number and the kinds of degrees of freedom that govern and delimit all human language development.

To understand how a "specialized ability" may simultaneously be a "corresponding restriction," we need only remind ourselves of a single-address machine. The fact that such a machine can decode a machine-language instruction in terms of a component (say, its eight leftmost bits) that is the instruction's operation code, and another component (its remaining bits) that is its address portion, implies at once that no other instruction format is, for that machine, admissible. Indeed, the very idea of the grammaticality of a whole computer program, let alone that of a single machine-language instruction, implies that there exist some symbol strings which, while they may superficially look like programs, are unintelligible and hence not admissible as programs. What is seen from one point of view as a specialized ability of a machine must be seen as a restriction from another perspective.

How then, Chomsky asks, can we gain an insight into the genetically endowed abilities that we call the mind? He answers that, given our present state of virtually total ignorance about the living brain, our best chance is to infer the innate properties of the mind from the highly restrictive principles of a "universal grammar." The linguist's first task is therefore to write grammars, that is, sets of rules, of particular languages, grammars capable of characterizing all and only the grammatically admissible sentences of those languages, and then to postulate principles from which crucial features of all such grammars can be deduced. That set of principles would then constitute a universal grammar. Chomsky's hypothesis is, to put it another way, that the rules of such a universal grammar would constitute a kind of projective description of important aspects of the human mind. He does not believe, of course, that people know these rules in the same way that they know, say, the rules of long division. Instead they know them (to use Polanyi's word) tacitly, that is, in the same way that people know how to maintain their balance while running. In both speaking and running, by the way, performance once mastered, deteriorates when an attempt is made to apply explicitly rules consciously.

In an important sense, then, Chomsky is one of our mythical explorers. Unable to inspect the insides of the found objects, human minds, and ignorant of whatever engineering principles may be relevant, e.g., the neurophysiology of the living brain, he sets out to infer the found object's laws from the evidence of its linguistic behavior.[1]

As far-reaching as the research aims of Chomsky's school are, they are modest compared to those of the leading scientists working in that branch of computer science called "artificial intelligence" (AI). Herbert A. Simon and Allen Newell, for example, together leaders of one of the most productive teams of AI researchers at Carnegie-Mellon University, Pittsburgh, claimed as early as 1958 that, in their own words;

> "There are now in the world machines that think, that learn and that create. Moreover, their ability to do these things is going to increase rapidly until—in the visible future—the range of problems they can handle will be coextensive with the range to which the human mind has been applied."[2]

They thus proclaimed the research aim of the new science, AI, to be nothing less than to build a machine whose linguistic behavior, to say the least, is to be equivalent to that of humans. Should AI realize this aim, it will have achieved the second, and very high indeed, level of understanding of human functions that we discussed for our explorers' understanding of the functions of the machines they encountered. In that context we fantasized that the explorers had succeeded in building a machine whose input-output behavior was, under any test whatever, indistinguishable from that of the machines they found, although the components of the two machine types need not have been the same.

In fact, the research goals of AI are much more ambitious than were those of our explorers, who intended only to understand how the machine they found generated its textual responses to the textual inputs it was given, whereas the goal of AI is to understand how an organism handles "a range of problems . . . coextensive with the range to which the human mind has been applied." Since the human mind has applied itself to, for example, problems of aesthetics involving touch, taste, vision, and hearing, AI will have to build machines that can feel, taste, see, and hear. Since the future in which machine thinking will range as widely as Simon and Newell claim it will is, at this writing, merely "visible" but not yet here, it is perhaps too early to speculate what sort of equipment machines will have to have in order to think about such human concerns as, say, disappointment in adolescent love. But there are machines today, principally at M.I.T., at Stanford University, and at the Stanford Research Institute, that have arms and hands whose movements are observed and coordinated by computer-controlled television eyes. Their hands have fingers which are equipped with pressure-sensitive pads to give them a sense of touch. And there are hundreds of machines that do routine (and even not so routine) chemical analyses, and that may therefore be said to have senses of taste. Machine production of fairly high-quality humanlike speech has been achieved, principally at M.I.T. and at the Bell Telephone Laboratories. The U.S. Department of Defense and the National Science Foundation are currently supporting considerable efforts toward the realization of machines that can understand human speech. Clearly, Simon's and Newell's ambition is taken seriously both by powerful U.S. government agencies and by a significant sector of the scientific community.

Given that individuals differ in their visual acuity, it is not to be expected that everyone even now can see the same future that was already visible to Simon and Newell in 1958. Nor is it necessary for psychologists to recognize the power of computer models of human functions in order to share Simon's and Newell's grandiose vision. Much humbler signs point the way, and even more directly.

Whatever else man is, and he is very much else, he is also a behaving organism. If man's understanding of himself is to be at least in part scientific, then science must be allowed to assume that at least some aspects of man's behavior obey laws that science can discover and formalize within some scientific conceptual framework. However naive and informal or, on the other hand, sophisticated and formal a notion of "information" one has in mind, it must be granted that man acts on (that is, responds to) information that impinges on him from his environment, and that his actions, especially his verbal behavior, inform his environment in turn. Whatever else man is, then, and again he is very much else, he is also a receiver and a transmitter of information. But even so, he is certainly more than a mere mirror that reflects more or less precisely whatever signals impinge on it; for he attends to only a small fraction of what William James called "the blooming, buzzing confusion" of sensations with which his environment bombards him, and he transforms that distillate of his world into memories, mental imagery of many sorts, speech and writing, strokes on piano keyboards, in short, into thought and behavior. Whatever else man is, then, and he is much else, he is also an information processor.

I will, in what follows, try to maintain the position that there is nothing wrong with viewing man as an information processor (or indeed as anything else) nor with attempting to understand him from that perspective, providing, however, that we never act as though any single perspective can comprehend the whole man. Seeing man as an information-processing system does not in itself dehumanize him, and may very well contribute to his humanity in that it may lead him to a deeper understanding of one specific aspect of his human nature. It could, for example, be enormously important for man's understanding his spirituality to know the limits of the explanatory power of an information- processing theory of man. In order for us to know those limits, the theory would, of course, have to be worked out in considerable detail.

Before we discuss what an information-processing theory of man might look like, I must say more about theories and especially about their relation to models. A theory is first of all a text, hence a concatenation of the symbols of some alphabet. But it is a symbolic construction in a deeper sense as well; the very terms that a theory employs are symbols which, to paraphrase Abraham Kaplan, grope for their denotation in the real world or else cease to be symbolic.[3] The words "grope for" are Kaplan's, and are a happy choice—for to say that symbols "find" their denotation in the real world would deny, or at least obscure, the fact that the symbolic terms of a theory can never be finally grounded in reality.

Definitions that define words in terms of other words leave those other words to be defined. In science generally, symbols are often defined in terms of operations. In physics, for example, mass is, informally speaking, that property of an object which determines its motion during collision with other objects. (If two objects moving at identical velocities come to rest when brought into head-on collision, it is said that they have the same mass.) This definition of mass permits us to design experiments involving certain operations whose outcomes "measure" the mass of objects. Momentum is defined as the product of the mass of an object and its velocity (mv), acceleration as the rate of change of velocity with time ($a = dv/dt$), and finally force as the product of mass and acceleration ($f = ma$). In a way it is wrong to say that force is "defined" by the equation $f = ma$. A more suitable definition given in some physics texts is that force is any influence capable of producing a change in the motion of a body.[4] The difference between the two senses of "definition" alluded to here illustrates that so-called operational definitions of a theory's terms provide a basis for the design of experiments and the discovery of general laws, but that these laws may then serve as implicit definitions of the terms occurring in them. These and still other problematic aspects of definition imply that all theoretic terms, hence all theories, must always be characterized by a certain openness. No term of a theory can ever be fully and finally understood. Indeed, to once more paraphrase Kaplan, it may not be possible to fix the content of a single concept or term in a sufficiently rich theory (about, say, human cognition) without assessing the truth of the whole theory.[5] This fact is of the greatest importance for any assessment of computer models of complex phenomena.

A theory is, of course, not merely any grammatically correct text that uses a set of terms somehow symbolically related to reality. It is a systematic aggregate of statements of laws. Its content, its very value as theory, lies at least as much in the structure of the interconnections that relate its laws to one another, as in the laws themselves. (Students sometimes prepare themselves for examinations in physics by memorizing lists of equations. They may well pass their examinations with the aid of such feats of memory, but it can hardly be said that they know physics, that, in other words, they command a theory.) A theory, at least a good one, is thus not merely a kind of data bank in which one can "look up" what would happen under such and such conditions. It is rather more like a map (an analogy Kaplan also makes) of a partially explored territory. Its function is often heuristic, that is, to guide the explorer in further discovery. The way theories make a difference in the world is thus not that they answer questions, but that they guide and stimulate intelligent search. And (again) there is no single "correct" map of a territory. An aerial photograph of an area serves a different heuristic function, say, for a land-use planner, than does a demographic map of the same area. One use of a theory, then, is that it prepares the conceptual categories within which the theoretician and the practitioner will ask his questions and design his experiments. *

Ordinarily, of course, when we speak of putting a theory to work, we mean drawing some consequences from it. And by that, in turn, we mean postulating some set of circumstances that involves some terms of the theory, and then asking what the theory says those particular circumstances imply for others of the theory's terms. We may describe the state of the economy of a specific country to an economist, for example, by giving him a set of the sorts of economic indices his particular economic theory accommodates. He may ask us some questions which, he would say, emerge directly from his theory. Such questions, by the way, might give us more insight into whether he is, say, a Marxist or a Keynesian economist than any answers he might ultimately give us, for they would reveal the structure of his theory, the network of connections between the eco-

---

* It must not be thought that this heuristic function of theory is manifest only in science. To name but one of the possible examples outside the sciences, Steven Marcus, the American literary critic, used theories of literary criticism freshly honed on the stone of psychoanalytic theory to do an essentially anthropological study of that "foreign, distinct, and exotic" subculture that was the sexual subculture of Victorian England. See his *The Other Victorians* (New York: Basic Books, 1966). More recently he wrote in the preface of his *Engels, Manchester, and the Working Class* (New York: Random House, 1974), "The present work may be regarded as part of a continuing experiment . . . to ascertain how far literary criticism can help us to understand history and society; to see how far the intellectual discipline that begins with the work of close textual analysis can help us understand certain social, historical, or theoretical documents." In neither book was a theory of literary criticism "applied," as, for example, a chemical theory may be applied to the chemical analysis of a compound; instead, Marcus' theories were used heuristically, as travelers use maps to explore a strange territory.

nomic laws in which he believes. Finally, we expect to be told what his theory says, e.g., that the country will do well, or that there will be a depression. More technically speaking, we may say that to put a theory to work means to assign specific values, by no means always numerical, to some of its parameters (that is, to the entities its terms signify), and then to methodically determine what values the theory assigns to other of its parameters. Often, of course, we arrive at the specifications to which we wish to apply a theory by interrogating or measuring some aspect of the real world. The input, so to speak, to a political theory may, for example, have been derived from public-opinion polls. At other times our specifications may be entirely hypothetical, as, for example, when we ask of physics what effect a long journey near the speed of light would have on the timekeeping property of a clock. In any case, we identify certain terms of the theory with what we understand them to denote, associate specifications with them, and, in effect, ask the theory to figure out the consequences.

Of course, a theory cannot "figure out" anything. It is, after all, merely a text. But we can very often build a model on the basis of a theory. And there are models which can, in an entirely nontrivial sense, figure things out. Here I am not referring to static scale models, like those made by architects to show clients what their finished buildings will look like. Nor do I mean even the scale models of wings that aerodynamicists subject to tests in wind tunnels; these are again static. However, the system consisting of both such a wing and the wind tunnel in which it is flown is a model of the kind I have in mind. Its crucial property is that it is itself capable of behaving in a way similar to the behaving system it represents, that is, a real airfoil moving in a real airmass. The behavior of the wing in the wind tunnel is presumably determined by the same aerodynamic laws as govern the behavior of the wings of real airplanes in flight. The aerodynamicist therefore hopes to learn something about a full-scale wing by studying its reduced-scale model.

The connection between a model and a theory is that a model *satisfies* a theory; that is, a model obeys those laws of behavior that a corresponding theory explicitly states or which may be derived from it. We may say, given a theory of a system *B*, that *A* is a model of *B* if that theory of *B* is a theory of *A* as well. We accept the condition also mentioned by Kaplan that there must be no causal connection between the model and the thing modelled; for if a model is to be used as an explanatory tool, then we must always be sure that any lessons we learn about a modeled entity by studying its model would still be valid if the model were removed.

People do, of course, derive consequences from theories without building explicit models like, say, scaled-down wings in wind tunnels. But that is not to say that they derive such consequences without building models at all. When a psychiatrist applies psychoanalytic theory to data supplied to him by his patient, he is, so to speak, exercising a mental model, perhaps a very intuitive one, of his patient, a model cast in psychoanalytic terms. To state it one way, the analyst finds the study of his mental model (*A*) of his patient (*B*) useful for understanding his patient (*B*). To state it another way, the analyst believes that psychoanalytic theory applies to his patient and therefore constructs a model of him in psychoanalytic terms, a model to which, of course, psychoanalytic theory also applies. He then transforms (translates is perhaps a better word) inferences derived from working with the model into inferences about the patient. (It has to be added, lest there be a misunderstanding, that however much the practicing psychoanalyst is committed to psychoanalytic theory and however much his attitudes are shaped by it, psychoanalytic therapy consists in only small part of direct or formal application of theory. Nevertheless, it is plausible that all of us make all our inferences about reality from mental models whose structures, and to a large extent whose contents as well, are strongly determined by our explicitly and implicitly held theories of the world.)

Computers make possible an entirely new relationship between theories and models. I have already said that theories are texts. Texts are written in a language. Computer languages are languages too, and theories may be written in them. Indeed, for the present purpose we need not restrict our attention to machine languages or even to the kinds of "higher-level" languages we have discussed. We may include all languages, specifically also natural languages, that computers may be able to interpret. The point is precisely that computers do *interpret* texts given to them, in other words, that texts determine computers' behavior. Theories written in the form of computer programs are ordinary theories as seen from one point of view. A physicist may, for example, communicate his theory of the pendulum either as a set of mathematical equations or as a computer program. In either case he will have to identify the terms of his theory—his "variables," in technical jargon—with whatever they are to correspond to in reality. (He may say l is the length of the pendulum's string, p its period of oscillation, g the acceleration due to gravity, and so on.) But the computer program has the advantage not only that it may be understood by anyone suitably trained in its language, just as a mathematical formulation can be readily understood by a physicist, but that it may also be run on a computer. Were it to be run with suitable assignments of values to its terms, the computer would *simulate* an actual pendulum. And inferences could be drawn from that simulation, and could be directly translated into inferences applicable to real pendulums. A theory written in the form of a computer program is thus both a theory and, when placed on a computer and run, a model to which the theory applies. Newell and Simon say about their information-processing theory of human problemsolving, "the theory performs the tasks it explains."[6] Strictly speaking, a theory cannot "perform" anything. But a model can, and therein lies the sense of their statement. We shall, however, have to return to the troublesome question of what the performance of a task can and cannot explain.

In order to aid our intuition about what it means for a computer model to "behave," let us briefly examine an exceedingly simple model: We know from physics, and indeed it follows from the equation $f = ma$ that we mentioned earlier, that the distance d an object will fall in a time t is given by

$$d = at^2/2,$$

where a is the acceleration due to gravity. In most elementary physics texts, a is simply asserted to be the earth's gravitational constant, namely, 32 ft/sec², where the unit of distance is feet and that of time is seconds. The equation itself is a simple mathematical model of a falling object. If we assume, for the sake of simplicity, that the acceleration *a* is indeed constant, namely, 32 ft/sec², we can compute how far an object will have fallen after, say, 4 seconds: $4 \times 4 = 16$ and $16 \times 32 = 512$ and $512 \div 2 = 256$. The answer, as schoolchildren would say, is therefore 256 feet.

  Mathematicians long ago fell into the habit of writing the so-called variables that appear in their equations as single letters. Perhaps they did this to guard against writer's cramp or to save chalk. Whatever their reasons, their notation is somewhat less than maximally mnemonic. Because computer programs are often intended to be read and understood by people, as well as to be executed by computers, and since computers are, within limits, indifferent to the lengths of the symbol strings they manipulate, computer programmers often use whole words to denote the variables that appear in their programs. Other considerations make it inconvenient to use juxtaposition of variables, as in *xy*, to indicate multiplication. Instead the symbol "∗" is used in many programming languages. Similarly, "∗∗" is used to indicate exponentiation. Thus, where the mathematician writes t^2, the programmer writes $t**2$. The equation

$$d = at^2/2$$

when transformed into a program statement* may thus appear as

distance = (acceleration ∗ time ∗∗2)/2.

  Let us now complicate our example just a little. Suppose an object is to be dropped from a stationary platform, say, a helicopter

---

* A significant technical point must be made here. Although the "statement" shown here is a transliteration of the equation to which it corresponds, it is not itself an equation. In technical parlance, it is an "assignment statement." It assigns a value to the variable "distance." "Distance," in turn, is technically an "identifier," the name of a storage location in which is stored the value which has been assigned to the corresponding variable. In mathematics, a variable is an entity whose value is not known, but which has a definite value nonetheless, a value that can be discovered by solving the equation. In programs, a variable may have different values at different stages of the execution of the program. In ordinary mathematics, e.g., in high-school algebra, the "equation" "$x = x + 1$" is nonsense. The same string of symbols appearing as an expression in a program has meaning, namely, that 1 is to be added to the contents of the location denoted by "x" and those contents replaced by the resulting sum.

hovering at some altitude above the ground. The object's height above the ground after it has fallen for some time would then be given by

$$\text{height} = \text{altitude} - (\text{acceleration} * \text{time} ** 2)/2.$$

Finally, suppose that the helicopter is flying forward at some constant velocity while maintaining its altitude. If there were no aerodynamic effects on the object dropped from the helicopter, it would remain exactly below the helicopter during its entire journey to the ground. The object's horizontal displacement from the point over which it was dropped would therefore be the same as the helicopter's horizontal displacement from that point, that is,

$$\text{displacement} = \text{velocity} * \text{time},$$

where by "velocity" we here, of course, mean the helicopter's velocity.

We now have, from one point of view, two equations, from another point of view, two program statements, from which we can compute the horizontal and vertical coordinates of an object dropped from a moving helicopter. We can combine them and imbed them in a small fragment of a computer program, as follows:

```
FOR time = 0 STEP .001 UNTIL height = 0 DO;
    height = altitude − (acceleration * time**2) / 2 ;
    displacement = velocity * time ;
    display (height, displacement) ;
END.
```

This is an example of a so-called *iteration statement.* It tells the computer to do a certain thing until some condition is achieved. In this case, it tells the computer to first set the variable "time" to zero, then to compute the height and displacement of what we would interpret to be the falling object, then to display the coordinates so computed—I shall say more about displaying in a moment—and, if the computed height is not zero, to add .001 to the variable "time" and do the whole thing again, that is, to iterate the process. (This program contains an error which, for the sake of simplicity, I have let stand. As it is, it may run forever. To repair it, the expression "height = 0" should be replaced by "height < 0." The reason for this is left to the reader to discover.)

We have assumed here that the computer on which this program is to run has a built-in display apparatus and the corresponding display instruction. We may imagine the computer's display to be a cathode-ray tube like that of an ordinary television set. The display instruction delivers two numbers to this device, in this example, the values of height and displacement. The display causes a point of light to appear on its screen at the place whose coordinates are determined by these two numbers, i.e., so many inches up and so many inches to the right of some fixed point of origin.

If we now make some additional assumptions about for example, the persistence of the lighted dot on the screen and the overall timing of the whole affair, we can imagine that the moving dot we see will appear to us like a film of the object falling from the helicopter (see Figure 5.1). It is thus possible, even compelling, to think of the computer "behaving," and for us to interpret its behavior as modeling that of the falling object.

It would be very easy for us to complicate our example step by step, first, for example, by extending it to cover the trajectory of a missile fired from a gun and, with that as a base, to extend it to the flight of orbiting satellites. We would then have described at least

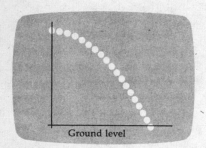

Ground level

Figure 5.1.
Cathode simulation of the trajectory of an object dropped from a flying helicopter.

the most fundamental basis on which the orbital simulations we often see on television are developed. But that is not my purpose. Simple as our example is, we can learn pertinent lessons from it.

To actually use the model, an investigator would initialize it by assigning values to the parameters altitude and velocity, run it on an appropriate computer, and observe its behavior on the computer's display device. There would, however, be discrepancies between what the model, so to speak, says a falling object would do and the behavior of its real counterpart. The model, for example, makes the implicit assumption that there are no aerodynamic effects on the falling object. But we know that there would certainly be air resistance in the real situation. Indeed, if the object dropped were a parachute, its passenger's life would depend on air resistance slowing its fall. A model is always a simplification, a kind of idealization of what it is intended to model.

The aim of a model is, of course, precisely not to reproduce reality in all its complexity. It is rather to capture in a vivid, often formal, way what is essential to understanding some aspect of its structure or behavior. The word "essential" as used in the above sentence is enormously significant, not to say problematical. It implies, first of all, purpose. In our example, we seek to understand how the object falls, and not, say, how it reflects sunlight in its descent or how deep a hole it would dig on impact if dropped from such and such a height. Were we interested in the latter, we would have to concern ourselves with the object's weight, its terminal velocity, and so on. We select, for inclusion in our model, those features of reality that we consider to be essential to our purpose. In complex situations like, say, modeling the growth, decay, and possible regeneration of a city, the very act of choosing what is essential and what is not must be at least in part an act of judgment, often political and cultural judgment. And that act must then necessarily be based on the modeler's intuitive mental model. Testing a model may reveal that something essential was left out of it. But again, judgment must be exercised to decide what the something might be, and whether it is "essential" for the purpose the model is intended to serve. The ultimate criteria, being based on intentions and purposes as they must be, are finally determined by the individual, that is, human, modeler.

The problem associated with the question of what is and what is not "essential" cuts the other way as well. A model is, after all, a different object from what it models. It therefore has properties not shared by its counterpart. The explorers we mentioned earlier may have built a functional model of the computer they found by using light-carrying fibers and light valves, whereas the real computer used wires and the kind of electronic gates we considered in Chapter III. They could then easily have come to believe that light is essential to the operation of computers. Their computer science might have included large elements of physical optics, and so on. It is indeed possible to build computers using light-carrying fibers, etc. Their logical diagrams, that is, their paper designs, would, up to a point, be indistinguishable from those of the corresponding electronic computers, because the former would have the same structure as the latter. What is essential about a computer is the organization of its components and not, again up to a point, precisely what those components are made of. Another example: there are people who believe it possible to build a computer model of the human brain on the neurological level. Such a model would, of course, be in principle describable in strictly mathematical terms. This might lead some people to believe that the language our nervous system uses must be the language of our mathematics. Such a belief would be an error of the kind we mean. John von Neumann, the great computer pioneer, touched briefly on this point himself:

> "When we talk mathematics, we may be discussing a *secondary* language, built on the *primary* language truly used by the central nervous system. Thus the outward forms of *our* mathematics are not absolutely relevant from the point of view of evaluating what the mathematical or logical language *truly* used by the central nervous system is."[7]

One function of a model is to test theories at their extreme limits. I have already mentioned that computers can generate films that model the behavior of a particle at extreme limits of relativistic velocities. Our own simple model of falling objects could be used in its present form to simulate, hence to calculate, the fall of an object from a spaceship flying near the surface of the moon. All we would have to do is to initialize acceleration to the number appropriate for the gravity existing on the moon's surface (providing, of course, that the spaceship is not so high above the surface of the moon that the effect of the moon's gravitational field would have been significantly changed—another implicit assumption). For that simulation exercise we would not have to have any components in our model corresponding to air resistance or other aerodynamic effects: the moon has no atmosphere. (Recall that an astronaut simultaneously dropped a feather and a hammer onto the moon's surface and that they both reached the ground at the same time.)

It is a fact, however, that the moon's gravitational field varies from place to place. These variations are thought to be due to so-called masscons, that is, concentrations of mass within the moon that act somewhat like huge magnets irregularly buried deep within the moon. The masscon hypothesis was advanced to account for observed irregularities in the trajectories of spacecraft orbiting the moon. It is, in effect, an elaboration of the falling-body model we have discussed. The elaborated model is the result of substituting a complex mathematical function (in other words, a subroutine) for the single term "acceleration" of our simple model. I mention it to illustrate the process, in this case properly applied, of elaborating a model to account for new and unanticipated observations. But the masscon elaboration was not the only possible extension of either the theory or its computer model. It could have been hypothesized, for example, that the moon is surrounded by a turbulent ether mantle whose waves and eddies caused the spaceship's irregular behavior. There are dozens of very good reasons for rejecting this hypothesis, of course, but a good programmer, given a lot of data, could more or less easily elaborate the model with which we started by adding "ether turbulence subroutines" so that, in the end, the model behaved just as the spaceship was observed to behave. Such a model would, of course, no longer look simple. Indeed, its very complexity, plus the precision to which it carried its calculations, might lend it a certain credibility.

Earlier I said that the value of a theory lies not so much in the aggregation of the laws it states as in the structure that interconnects them. The trouble with the kind of model elaboration that would result from such an "ether turbulence" hypothesis is that it simply patches one more "explanation" onto an already existing structure. It is a patch in that it has no roots in anything already present in the structure. Computer models have, as we have seen, some advantages over theories stated in natural language. But the latter have the advantage that patching is hard to conceal. If a theory written in natural language is, in fact, a set of patches and patches on patches, its lack of structure will be evident in its very composition. Although a computer program similarly constructed may reveal its impoverished structure to a trained reader, this kind of fault cannot be so easily seen in the program's performance. A program's performance, therefore, does not alone constitute an adequate validation of it as theory.

I have already alluded to the heuristic function of theories. Since models in computer-program form are also theories (at least, some programs deserve to be so thought of), what I have said about theories in general also applies to them, perhaps even more strongly, in this sense: in order for us to draw consequences from discursive theories, even to determine their coherence and consistency, they must, as I have said, be modeled anyway, that is, be modeled in the mind. The very eloquence of their statements, especially in the eyes of their authors, may give them a persuasive power they hardly deserve. Besides, much time may elapse between the formulation of a theory and its testing in the minds of men. Computer programs tend to reveal their errors, especially their lack of consistency, quickly and sharply. And, in skilled hands, computer modeling provides a quick feedback that can have a truly therapeutic effect precisely because of its immediacy. Computer modeling is thus somewhat like Polaroid photography: it is hard to maintain the belief that one has taken a great photograph when the counterexample is in one's hands. As Patrick Suppes remarked,

The attempt to characterize exactly models of an empirical theory almost inevitably yields a more precise and clearer understand-

ing of the exact character of a theory. The emptiness and shallow-
ness of many classical theories in the social sciences is well brought
out by the attempt to formulate in any exact fashion what consi-
tutes a model of the theory. The kind of theory which mainly
consists of insightful remarks and heuristic slogans will not be
amenable to this treatment. The effort to make it exact will at the
same time reveal the weakness of the theory."[8]

The question is, of course, just what kinds of theories are "amenable
to this treatment?"

6

COMPUTER MODELS IN PSYCHOLOGY

Sometimes a very complex idea enters the public consciousness in a form so highly simplified that it is little more than a caricature of the original; yet this mere sketch of the original idea may nevertheless change the popular conception of reality dramatically. For example, consider Einstein's theory of relativity. Just how and why this highly abstract mathematical theory attracted the attention of the general public at all, let alone why it became for a time virtually a public mania and its author a pop-culture hero, will probably never be understood. But the same public which clung to the myth that only five people in the world could understand the theory, and which thus acknowledged its awe of it, also saw the theory as providing a new basis for cultural pluralism; after all, science had now established that everything is relative. A more recent example may be found in the popular reception of the work of F. Crick and J. D.

Watson, who shared the Nobel prize in Medicine in 1962 for their studies of the molecular structure of DNA, the nucleic acid within the living cell that transmits the hereditary pattern. Here again highly technical results, reported in a language not at all comprehensible to the layman, were grossly oversimplified and overgeneralized in the public mind into the now-popular inpression that it is already possible to design a human being to specifications decided on in advance. In one fell swoop, the general public created for itself a vision of a positive eugenics based not on such primitive and (I hope) abhorrent techniques as the killing and sterilization of "defectives," but on the creation of supermen by technological means. What these two examples have in common is that both have introduced new metaphors into the common wisdom.

A metaphor is, in the words of I. A. Richards, "fundamentally a borrowing between and intercourse of thoughts, a transaction between contexts."[1] Often the heuristic value of a metaphor is not that it expresses a new idea, which it may or may not do, but that it encourages the transfer of insights, derived from one of its contexts, into its other context. Its function thus closely resembles that of a model. A Western student of Asian societies may, for example, not learn anything directly from the metaphoric observation that the overseas Chinese are the Jews of Asia. But it may never have occurred to him that the position of Jews in the Western world, e.g., as entrepreneurs, intellectuals, and targets of persecution, may serve as a model that can provoke insights and questions relevant for understanding the social role and function of, say, the Chinese in Indonesia. Although calling that possibility to his attention may not give the Western student a new idea, it may enable him to derive new ideas from the interchange of two contexts, neither of which are themselves new to him, but which he had never before connected.

Neither the idea of one object moving relative to another, nor that of man being fundamentally a physical object, was new to the common wisdom in the 1920's and the 1960's, respectively. What struck the popular imagination when, for some reason, the press campaigned for Einstein's theory, was that science appeared to have pronounced relativity to be a fundamental and universal fact. Hence the slogan "everything is relative" was converted into a legitimate contextual framework which could, potentially at least, be coupled to every other universe of discourse, e.g., as an explanatory model legitimating cultural pluralism.

The results announced by Crick and Watson fell on a soil already prepared by the public's vague understanding of computers, computer circuitry, and information theory (with its emphasis on codes and coding), and, of course, by its somewhat more accurate understanding of Mendelian genetics, inheritance of traits, and so on. Hence it was easy for the public to see the "cracking" of the genetic code as an unraveling of a computer program, and the discovery of the double-helix structure of the DNA molecule as an explication of a computer's basic wiring diagram. The coupling of such a conceptual framework to one that sees man as a physical object virtually compels the conclusion that man may be designed and engineered to specification.

There is no point in complaining that Einstein never intended his theory to serve as one half of the metaphor just described. It is, after all, necessary for the two contexts coupled by a metaphor to be initially disjoint, just as (as I insisted earlier) a model must not have a causal connection to what it models. The trouble with the two metaphoric usages we have cited is that, in both, the metaphors are overextended. Einstein meant to say that there is no fixed, absolute spacetime frame within which physical events play out their destinies. Hence every description of a physical event (and, in that sense, of anything) must be relative to some specified spacetime frame. To jump from that to "everything is relative" is to play too much with words. Einstein's contribution was to demonstrate that, contrary to what had until then been believed, motion is not absolute. When one deduces from Einstein's theory that, say, wealth and poverty are relative, in that it is not the absolute magnitudes of the incomes of the rich and poor that matters, but the ratios of one to the other, one has illicitly elevated a metaphor to the status of a scientific deduction.

The example from molecular biology illustrates an overextension of a metaphor in another sense; there the extent of what we know about the human as a biological organism is vastly exaggerated. The result is, to say the least, a premature closure of ideas. The metaphor, in other words, suggests the belief that everything that needs to be known is known.

The computer has become a source of truly powerful and often useful metaphors. Curiously, just as with the examples already cited, the public embrace of the computer metaphor rests on only the vaguest understanding of a difficult and complex scientific concept (here, the theory of computability and the results of Turing and Church concerning the universality of certain computing schemes). The public vaguely understands—but is nonetheless firmly convinced—that any effective procedure can, in principle, be carried out by a computer. Since man, nature, and even society carry out procedures that are surely "effective" in one way or another, it follows that a computer can at least imitate man, nature, and society in all their procedural aspects. Hence everything (that word again!) is at least potentially understandable in terms of computer models and metaphors. Indeed, on the basis of this unwarranted generalization of the words "effective" and "procedure," the word "understanding" is also redefined. To those fully in the grip of the computer metaphor, to understand X is to be able to write a computer program that realizes X. This is vividly exemplified by Professor Marvin Minsky, Director of M.I.T.'s Artificial Intelligence Laboratory, who writes,

> [For computers] "to write really good music or draw highly meaningful pictures will of course require better *semantic* models in these *areas*. That these are not available is not so much a reflection on the state of heuristic [computer] programs as on the traditionally disgraceful state of analytic criticism in the arts—a cultural consequence of the fact that most esthetic analysts wax indignant when it is suggested that it might be possible to understand what they are trying to understand."[2]

Clearly, what Minsky means by "understanding" music or painting is quite different from what, say, Mozart or Picasso meant by the same term. One of his uses of the word "understand" in this quoted passage is essentially a pun—though, I belive, an unconscious one—

on the other. The very innocence of his use of it testifies to the tenacity of the hold the metaphor has on him. *

This new definition of understanding is now very widely accepted, not only explicitly in scientific circles, but implicitly in the common wisdom. It implies, as the psychologist George A. Miller ruefully pointed out,

> "that the only reason something cannot be done by a universal Turing machine is that we don't understand it. Given this interpretation of what 'understanding' consists of, any attempt to suggest counterexamples becomes merely a confession of ignorance or, if one persists in claiming that he can understand something he cannot describe explicitly, one becomes a prototypical member of that class of people known as mystics."[3]

In other words, the computer metaphor has become another lamppost under whose light, and only under whose light, men seek answers to burning questions.

No branch of science has erected this lamppost more deliberately and with more enthusiasm than has psychology. George Miller writes:

> "Many psychologists have come to take for granted in recent years . . . that men and computers are merely two different species of a more abstract genus called 'information processing systems.' The concepts that describe abstract information processing systems must, perforce, describe any particular examples of such systems."[4]

The narrowing of vision that characterizes modern scientific investigation, just like the narrowing of the field of view accomplished by a microscope, can only be justified, and sustained, because it permits us to see things we could otherwise not see. Science and technology have, after all, momentous achievements to their credit. Given the depth to which the computer metaphor has pene-

---

* It must be said that Prof. Minsky did not adopt the computer metaphor because of a naive misunderstanding of the theory of computability and its implications. To the contrary, as a deservedly acknowledged authority in computer science, he adopted the metaphor thoughtfully and deliberately.

trated psychology, it is natural to ask whether it has justified itself there in terms of some tangible achievements.

The impact that the computer, in its role as a high-speed numerical calculator, has had on psychology, although undoubtedly very large, hardly counts, as a "tangible achievement." Psychology has long tried to become "scientific" by imitating that most spectacularly successful science, physics. Psychologists, however, seemed for a very long time to have misunderstood just what it was that made physics somehow more a science than psychology. Like sociology too, psychology mistook the most superficial property of physics, its apparent preoccupation with numbers and mathematical formulas, for the core that makes it a science. Large sections of psychology therefore tried to become as mathematical as possible, to count, to quantify, to identify its numbers with variables (preferably ones having subscripted Greek letters), and to manipulate its newfound variables in systems of equations (preferably differential equations) and in matrices just as the physicists do. The very profusion of energy expended on this program was bound to guarantee that some useful results would be achieved. Psychometrics, for example, is and remains an honorable trade. And there can be no question that statistics benefited enormously from the exercise it was given by psychology, just as it had benefited in the days of its infancy from its application to gambling. Perhaps it repaid its two patrons about equally.

It is often said that the computer is merely a tool. The function of the word "merely" in that statement is to invite the inference that the computer can't be very important in any fundamental sense because tools themselves are not very important. I have argued that tools shape man's imaginative reconstruction of reality and therefore instruct man about his own identity. Yet the folk wisdom that perceives the computer as a basically trivial instrument rests on an accurate insight: the computer, used as a "number-cruncher" (that is, merely as a fast numerical calculator, and it is so used especially in the behavioral sciences), has often, as George Miller has also pointed out, put muscles on analytic techniques that are more powerful than the ideas those techniques enable one to explore. "The methodological rigorists," writes Stanislav Andreski, "are like cooks who would show us all their shiny stoves, mixers, liquidizers and what not, without ever making anything worth eating."[5] The high-speed number-cruncher is, in the hands of many psychologists, merely their newest, shiniest, and most spectacular mixer-liquidizer.

When the computer is used merely as a numerical tool in psychology (or in any other field), it does not usually create a focusing of vision; i.e., it is not comparable to the microscope. It is therefore unrealistic to expect such use to uncover previously unseen worlds or to render distinct what was earlier seen only in vague outline. Because the view of man as a species of the more general genus "information-processing system" does concentrate our attention on one aspect of man, it invites us to cast all his other aspects into the darkness beyond what that view itself illuminates. We are entitled to ask what we would purchase at that cost.

There can be no final answer to such a question, for the extent of the creative analogical reach of a metaphor must, by its very nature, be always surprising and thus not fathomable in advance. We can say in anticipation that the power of a metaphor to yield new insights, depends largely on the richness of the contextual frameworks it fuses, on their potential mutual resonance. How far that potential will be realized depends, of course, on how profoundly the participants in the creative metaphoric act can command both contexts. That is why, for example, the computer expert who knows nothing but computers (the "Fach Idiot," as the Germans call such a person) can derive no broad intellectual nourishment from his expertise and is therefore doomed to remain forever a hacker. That is also why the computer metaphor is, as George Miller puts it, "most productive in areas where a considerable foundation of theory based on previous research already exists."[6]

One area of psychology was extraordinarily well-prepared to benefit from a fusion with the kind of process-oriented thinking characteristic of computer scientists; it was the area which concerns itself with the cognitive processes underlying the acquisition and memorization of information. An enormous amount of laboratory work had been done on, for example, the task of memorizing so-called nonsense syllables. One form of an experiment that has been performed by countless psychological laboratories is to present a subject with, say, a dozen pairs of three-letter syllables, one pair at a time, and to ask him, on each (but the first) presentation of the first of the pair, to say what the second is. The syllables are carefully chosen to be inherently meaningless. Thus, for example, CAT is not a nonsense syllable, but PAG is. Subjects are exposed to the list, one pair at a time, repeatedly until they are able to give the correct response item to each stimulus item. Edward S. Feigenbaum reported,

"The phenomena of rote learning are well-studied, stable, and reproducible. For example, in the typical behavioral output of a subject, one finds:
  a. Failures to respond to a stimulus are more numerous than overt errors.
  b. Overt errors are generally attributable to confusion by the subject between similar stimuli or similar responses.
  c. Associations which are given correctly over a number of trials sometimes are then forgotten, only to reappear and later disappear again. This phenomenon has been called oscillation.
  d. If a list *x* of syllables or syllable pairs is learned to the criterion; then a list *y* is similarly learned; and finally retention of list *x* is tested; the subject's ability to give the correct *x* responses is degraded by the interpolated learning. The degradation is called retroactive inhibition. The overt errors made in the retest trial are generally intrusions from the list *y*. The phenomenon disappears rapidly. Usually after the first retest trial, list *x* has been relearned back to criterion.
  e. As one makes the stimulus syllables more and more similar, learning takes more trials."[7]

Feigenbaum, currently a professor of computer science at Stanford University, conjectured that this sort of learning task involved the subject in an active, complex symbol-manipulation process which could best be described and understood in terms of more elementary symbol-manipulation processes of just the kind that can be programmed for a computer.

Of course, nothing would have been easier than to write a small program for a computer which would have enabled an experimenter to give the computer lists of nonsense syllables that the computer could then reproduce perfectly after the first "trial." The task Feigenbaum set for himself was much harder: to produce, in the form of a computer program, a model of cognitive processes whose over-all behavior would closely approximate that of human subjects engaged in memorizing nonsense syllables, and whose detailed internal functions would constitute a theoretical explanation of the difficulties actually observed in experiments. Moreover, he wished his explanations to be at least consistent with such psychological observations as, for example, that humans have both short-term memories, in which they can apparently hold a few symbols for instant recall during a short period of time, and longer-term memories, in which an almost unlimited amount of information can be stored but from which individual items can be retrieved only at the expense of some effort. If this "effort" to remember is thought of as the computational effort involved in executing a perhaps long subroutine, it becomes obvious how one can begin to apply the computer metaphor.

Feigenbaum's central idea is for the computer to store *descriptions* of the syllables presented to it, not the actual syllables themselves. The syllable DAX, for example, may be described by the fact that its first letter has a vertical leading edge and contains a closed loop, that its second letter contains a horizontal middle bar, and so on. When a syllable is first presented to the system, a description of it just sufficiently detailed to allow it to be discriminated from the syllables already stored is added to storage. If it is a stimulus item, that is, the first of a syllable pair, then a "cue" consisting of a minimal description of the syllable with which it is to be associated is stored with it. Because all these descriptions are so minimal, the system often makes wrong associations when presented with stimulus items. But because the correct response item is presented whenever the system makes such an error, the descriptive information then in play may be improved by adding further description to it. Eventually the system learns the list perfectly. When another list is then attempted, the descriptions associated with it may again be confused with those corresponding to the first list, and vice versa. The system is thus capable of exhibiting retroactive inhibition. And clearly, as the items to be learned are made more and more similar to one another, an increasing number of trials is required to refine the discriminating power of each relevant descriptor. The system thus behaves very much as does a human confronted with the same task.

Feigenbaum's program, though by now very old (it was completed in 1959), remains instructive in at least two respects. First, it offers us a relatively simple example of what is meant by a model of a cognitive process in computer-program form. The way it organizes its information storage is meant to be a functional description of the human intermediate-term memory. As such, it explains, for example, how it may be that we can totally forget something for a long time and yet recall it again under certain circumstances. It cannot be that the allegedly forgotten item was simply wiped out of our minds, for if it were, we could never regain access to it. In Feigenbaum's model no information is ever destroyed. But information may be hidden, so to speak, by descriptors leading to other associations; thus one memory may screen or mask another. Sometimes a refinement of the screening image (that is, of a cue) is, in Feigenbaum's system, all that is required to uncover (that is, to make again accessible) what was previously masked.

Feigenbaum's system also requires that the two syllables to be associated with one another be simultaneously available to the computer (that is, present in its store) for a short time. After a "cue" to the response item has been associated with the description of the stimulus syllable, the two syllables per se can be erased from the computer's store—in other words "forgotten." There is thus a part of his system that plausibly corresponds to what little psychologists know about the function of the human short-term memory. No one, least of all Feigenbaum, claims that his model constitutes "the" explanation for such phenomena. But it is an explanation in a domain where explanations are rare.

The second respect in which Feigenbaum's program is instructive is that it behaves in ways which were not directly and deliberately "programmed in," as the saying goes. For example, the program exhibits what psychologists call interference; that is, the acquisition of a new association interferes with the production of an older one when the syllables involved have closely similar descriptions. The program contains no interference subroutine as such. The phenomenon arises as a consequence of the entire structure of the program, and appeared as a surprise to its designer. In that respect, then, the model *predicted* a behavioral phenomenon, which enormously enhanced its plausibility. The program thus instructs us that the easy and much-repeated slogan "a computer does only what its programmer told it to do" is in certain respects quite wrong and is in any case problematical.

The program we have been discussing is a member of a class of programs called "simulation programs." Their object is to simulate the way humans accomplish certain tasks, but decidedly not to accomplish those tasks in the most efficient way a computer possibly could. We have noted, for example, that a computer could easily be programmed to "memorize" lists of nonsense syllables in one "trial." But that would teach us nothing about how humans might accomplish what appears at least superficially to be the same task.

Because programs which concern the cognitive aspects of human behavior fall naturally within the domain of artificial intelligence, AI (about which more later), they need be distinguished from another class of AI programs, namely, ones that are entirely task-oriented.

Workers in AI tend to think of themselves as working in one of two modes, often called *performance mode* and *simulation mode.* Perhaps the best way to make the distinction clear is by analogy to flying. Virtually all early attempts to understand flying or to build flying models were based on imitating the flight of birds. It is a plausible conjecture that the myth of Icarus, the Greek hero who flew with wings attached to his body by wax and who crashed when the heat of the sun melted the wax, reflects man's early failure to imitate the birds. We might say that the early thinkers and pioneers were operating in simulation mode. Already by the middle of the nineteenth century, however, men like Henson and Stringfellow, and somewhat later Langley, shifted to what we might call performance mode. They considered that their task was to build flying machines based on whatever principles they could discover. Their aim was performance first and understanding only to the extent that it would contribute to performance.

A third mode of operation should perhaps be mentioned in this context: theory mode. There were great aerodynamicists before there were practical aircraft. Lord Rayleigh, for example, published important papers specifically on the theory of flight beginning about 1875. Of course, after the Wright brothers achieved their historic flights in 1903, interest in aerodynamics increased continually and has not flagged to this day. But whereas the aerodynamicist is devoted to theory as such and tends to think of practical aircraft as being mere models of his theories, the aircraft designer looks to theory as being merely another source of ideas which may help him gain more performance from his machines.

The situation in AI is closely analogous to that just described. The goal of a majority of workers in AI is to build machines that behave intelligently, whether or not what they produce sheds any light on human intelligence. They are working in performance mode. They wish to build machines that speak as humans do and that understand human speech, that can, with the aid of television eyes and mechanical arms and hands, screw nuts onto bolts and assemble even more complex mechanical gadgets, that can analyze and synthesize chemical compounds, that can translate natural languages from one to another, that can compose music and complex computer programs, and so on. They are, of course, happy to accept whatever contributions the theoreticians (for example, the psychologists) can make that may help them realize their wishes. But their goal, unlike that of the theoretician who seeks understanding (or claims to), is performance first and last.

A program like Feigenbaum's clearly eschews performance; it is designed to require many trials to learn its lists, whereas, as I have pointed out, if performance were its goal, it could be made to memorize them in one trial.

The dividing line between simulation mode and performance mode is, as might be expected, not absolute. Often the only way to begin thinking about how to get a computer to do a specific task is to ask how people would do it. One thus starts out essentially simulating one's own introspectively observed behavior.

There is, of course, a difference between a program whose avowed aim is performance but that, at least initially, simulates "the way people do it," and a program that simulates what people do in order to learn something about people. But when a program undertaken under the latter banner succeeds, it also performs. Sometimes its authors then cannot resist the temptation to make performance as well as theoretical claims for it, and thus to contribute to the blurring of the line dividing performance mode from simulation mode. Newell, Shaw, and Simon, for example, wrote a program which could prove some theorems in the propositional calculus by simulating the way students who are naive about logic struggled with such proofs.[8] Newell, Shaw, and Simon stated as their aim, "we wish to understand how a mathematician, for example, is able to prove a theorem even though he does not know when he starts how, or if, he is going to succeed." After reporting how long it took their program to prove a number of theorems, they remarked: "One can invent 'automatic' procedures for producing proofs . . . but these turn out to require computing times of the order of thousands of years for the proof of [some particular theorem]." It is hard to read that statement as anything other than a claim that their program can perform usefully, aside from its being possibly instructive about how "mathematicians prove a theorem." As it happened, within a year or two after the appearance of their paper, the mathematician Hao Wang published an "automatic procedure," that is, a computer program, capable of proving all theorems in the propositional calculus. It proved the particular theorem whose proof Newell, Shaw, and Simon estimated would "require computing times of the order of thousands of years" in 1/4 second on what today would be considered a very primitive computer.[9]

The fuzziness of the line dividing simulation made from performance mode is, quite justifiably, a matter of little concern to workers in AI. At the outset of a large research effort, what is important is to have a fairly clear idea of at least the general domain within which questions are to be asked, or, to put it another way, of what it is that is not presently understood that the research is intended to help us understand. Wang's research yielded a result that deepened our understanding of certain aspects of mathematical logic. The aim of the work reported by Newell, Shaw, and Simon was, in their own words, to "understand the complex processes (heuristics) that are effective in problem solving."[10] They chose to examine how people prove theorems merely as an example of human problem solving. Newell and Simon have, as we shall see, pursued their work on problem solving to this very day. What has been problematical about it, and remains so, is in what sense of the word "understand" it helps us to understand man as an information processor or as anything else. That same question can usefully be asked about artificial intelligence generally.

That I have so far cited only very early AI projects is not in any sense "unfair," for AI researchers themselves continue to cite these very examples (namely, Feigenbaum's program, and the logic theory machine of Newell, Shaw, and Simon) as being fundamental work, as far as they go. Newell and Simon have, as I have said, continued their work on problem solving. Meanwhile other workers, notably those at M.I.T.'s and Stanford University's Artificial Intelligence Laboratories have increasingly turned their attention to robotics, that is, to the problems associated with the building of machines that sense aspects of their environments, e.g., with the aid of television eyes, and that are capable physically acting on it, e.g., by means of computer-controlled mechanical arms and hands. Their work has, as might be expected, generated a host of subproblems in such areas as vision, computer understanding of natural language, and pattern recognition.

Of course, some of these subproblems are also autonomous problems, that is, are independent of the research goals of robotics. Natural-language understanding by computer is an example of a problem that is inherently interesting and difficult in its own right. That any progress on it may prove useful for instructing robots is, to many workers, merely an additional motivation, certainly not the principal one. I shall later have more to say about the manipulation of natural language by computers. For the moment, however, let us turn to some of the more recent work on problem solving, particularly that of Newell and Simon.

The modern literature on problem solving is punctuated by two important books, George Polya's *How to Solve It* and Newell's and Simon's *Human Problem Solving*.[11] Polya's book was first published in 1945, that is, years before electronic computers became practical research instruments. Yet in it Polya lays the groundwork and, in a sense, heralds all the work on problem solving that was to follow for thirty years afterward. Polya's concern is with heuristic problem-solving methods, that is, with those rules of thumb which, when applied, may well lead to a solution of the problem at hand or to some progress toward solving it, but which do not guarantee a solution. Heuristics are thus not algorithms, not effective procedures; they are plausible ways of attacking specific problems. Polya anticipated much of the later work of computer scientists on problem solving when he wrote:

> "Modern heuristic endeavors to understand the process of solving problems, especially the *mental operations typically useful in this process*. . . . Experience in solving problems and experience in watching other people solving problems must be the basis on which heuristic is built. In this study, we should not neglect any sort of problem, and should find out common features in the way of handling all sorts of problems; we should aim at general features, independent of the subject matter of the problem. The study of heuristic has 'practical' aims; a better understanding of the mental operations typically useful in solving problems. . . .
>
> "It is emphasized that all sorts of problems, especially PRACTICAL PROBLEMS, and even PUZZLES, are within the scope of heuristic [sic]. It is also emphasized that infallible RULES OF DISCOVERY are beyond the scope of serious research. Heuristic discusses human behavior in the face of problems. . . . Heuristic aims at generality, at the study of procedures which are independent of the subject-matter and apply to all sorts of problems."[12]

No clearer charter for the work of Newell and Simon could have been written. Polya, in effect, predicted those aspects of what Newell and Simon were later to do which most truly characterize their conception: the endeavor to understand mental operations, the emphasis on generality, on independence from subject matter, and on the usefulness of watching people solve problems, and the stress laid on puzzle-solving behavior. Finally, Polya emphasizes that his book is about methods, and that the most important heuristic is "the end suggests the means."

What Newell and Simon were later to call "the means-ends method" was first suggested when the way an early version of their logic-theory machine proved theorems was compared with recordings of "thinking aloud" sessions of nonmathematics students attempting the same tasks. These so-called *protocols* proved highly suggestive for further work. Protocol taking, that is, watching other people solve problems, became virtually a hallmark of Newell and Simon's procedure.

The new information-processing psychology proceeds from the basic view

> "that programmed computer and human problem solver are both species belonging to the genus 'Information Processing System' (IPS). . . .
>
> "When we seek to explain the behavior of human problem solvers (or computers for that matter), we discover that their flexibility—their programmability—is the key to understanding them. Their viability depends upon their being able to behave adaptively in a wide range of environments. . . .
>
> "If we carefully factor out the influences of the task environments from the influences of the underlying hardware components and organization, we reveal the true simplicity of the adaptive system. For, as we have seen, we need postulate only a very simple information processing system in order to account for human problem solving in such tasks as chess, logic, and cryptarithmetic. The apparently complex behavior of the information processing system in a given environment is produced by the interaction of the demands of that environment with a few basic parameters of system, particularly characteristics of its memories.
>
> "Matters are simple, not because the law of large numbers cancels things out, but because things line up in a means-ends chain in which only the end points count (i.e., equifinality)."[13]

This is a truly remarkable statement, especially in light of Simon's claims that the hypothesis it represents "holds even for the whole man." It behooves us to attempt to understand just what this "very simple" information-processing system is which produces complex behavior as a function of its environment and "a few basic parameters." We must also ask what it is about tasks like chess, logic, and cryptarithmetic that generalizes to the "whole range [of problems] to which the human mind has been applied," that is, to that range to which these same authors have promised computers will be applied "in the visible future."[14] The last question is especially pertinent because existing heuristic problem-solving programs deal only with very simple problems in chess and logic. Cryptarithmetic hardly counts here, since it is what is called, even in AI circles, a "toy problem."*

We have already agreed that it is entirely proper and even useful to assume a very particular viewpoint and, from the perspective it affords, to see man as an information processor. And since the computer, the Turing machine, is a universal information processor, it is natural to compare man as seen from that perspective with the computer. Information-processing *psychology* is, however, *not* information-processing *neurophysiology*. It does not attempt explanations in terms of bits or by making analogies to flip-flops, electronic circuits, and so on. It eschews, and rightly so, even explanations that depend on comparison with the sort of symbol manipulations that classical Turing machines do, e.g., writing, reading, erasing, and comparing extremely simple and irreducible symbols such as zeros and ones.

Recall that a program for a particular computer is essentially a description of another computer, that it transforms the former machine into the latter. One can therefore design a computer, and subsequently implement it in the form of a computer program whose "built-in" elementary information processes (*eip's*—the terminology is Newell and Simon's) are ones that operate on arbitrarily

---

* True, there are very powerful programs for doing extremely complex symbolic logic and mathematics. But these are special-purpose programs which—although they may have benefited from AI techniques in their early development—can in no way be seen as the kinds of information-processing system Newell and Simon talk about. They are, beyond dispute, no more relevant to psychology than are the many programs which solve systems of differential equations. The currently most powerful chess programs were also written in performance mode, and, although they may use certain of the techniques of the AI armamentarium, they too are essentially irrelevant to psychology.

complex symbol structures, that is, that read, write, erase, compare such symbol structures with one another, and so on. Such structures can be made to represent formulas in logic, mathematical expressions, words, sentences, architectural drawings, etc., and, of course, computer programs which may themselves then be manipulated by eip's.

A particularly useful programming device is, for example, to organize information in the form of concatenations of individual items. The "link" that chains one item to the next is a machine address, a pointer, that is stored next to one of the items and points to its successor. A list (as such chains are commonly called) of items so concatenated may then again be considered an item and may thus be pointed to by still another item. In this way, structures of very great complexity may be created and manipulated. Specific eip's may treat them as single items, whereas others may course over them, inserting and deleting substructures, for example.

An information-processing system is therefore, in this context, a hardware computing system together with a program capable of executing eip's on stored symbol structures. It has, of course, input-output equipment, such as console typewriters, that enable adequate communication with the world outside itself.

The most ambitious information-processing system that has been built for the purpose of studying human problem-solving behavior is Newell and Simon's General Problem Solver (GPS).[15]

"The main methods of GPS jointly embody the heuristic of means-ends analysis. . . . Means-ends analysis is typified by the following kind of common-sense argument:

I want to take my son to nursery school. What's the difference between what I have and what I want? One of distance. What changes distance? My automobile. My automobile won't work. What is needed to make it work? A new battery. What has new batteries? An auto repair shop. I want the repair shop to put in a new battery; but the shop doesn't know I need one. What is the difficulty? One of communication. What allows communication? A telephone . . . and so on.

This kind of analysis—classifying things in terms of the functions they serve, and oscillating among ends, functions required, and means to perform them—forms the basic system of heuristic of GPS. More precisely, this means-ends system of heuristic assumes the following:

1. If an object is given that is not the desired one, differences will be detectable between the available object and the desired object.

2. Operators affect some features of their operands and leave others unchanged. Hence operators can be characterized by the changes they produce and can be used to try to eliminate differences between the objects to which they are applied and desired objects.

3. If a desired operator is not applicable, it may be profitable to modify its inputs so that it becomes applicable.

4. Some differences will prove more difficult to affect than others. It is profitable, therefore, to try to eliminate 'difficult' differences, even at the cost of introducing new differences of lesser difficulty. This process can be repeated as long as progress is being made toward eliminating the more difficult differences."[16]

To see how this works on one of the kinds of problems to which GPS has actually been applied, consider the following cryptarithmetic puzzle:

$$\begin{array}{r} DO \\ + \quad IT \\ \hline TTD \end{array}$$

A subject is told that the above is an encoding of a problem in ordinary addition. Each letter represents a number, and no two letters represent the same number. His task is to assign numbers to the

letters in such a way that the given expression represents a correct
addition. He is to produce a protocol, that is, to say out loud what he
is thinking. Following is one possible such protocol, interspersed
with an analysis in **GPS**-like terms:

> *Subject:* $D + I$ must be greater than 9 because there is a
> carry to the next column.
>
> *Analysis:* The subject applied the operator "process col-
> umn."
>
> *Subject:* T must be 1 since it is a carry.
>
> *Analysis:* The subject applied the operator "assign value."
> He has reached a subgoal and reduced the difference
> between the given and the desired object. The "given
> object" is now

$$
\begin{array}{r}
D\,O \\
+\ \ I\ 1 \\
\hline
1\,1\ D
\end{array}
$$

Subject: O must be at least 2.

Analysis: The subject applied the operator "generate possible values" to O. (There must have been some un-
spoken tentative application of the operator "assign
value" whose results were rejected.)

Subject: Let's try $O = 2$.

Analysis: The subject applied the operator "assign value."
Another reduction of difference. The "given object" is
now

$$
\begin{array}{r}
3\,2 \\
+\ \ I\ 1 \\
\hline
1\,1\,3
\end{array}
$$

Subject: $I = 8$.

Analysis: The "assign value" operator is applied and the
difference between the given object and the desired
object removed. The goal is reached.

This is a much simpler problem than those typically given to subjects and to GPS. A much more typical example of a problem that has been fully analyzed is

$$DONALD$$
$$+ GERALD$$
$$\overline{ROBERT,}$$

where $D = 5$. The example we have worked out suffers from the additional fault that it does not display any wrong moves, backtracking, and so on. Nevertheless it gives a general, if pale, idea of the way GPS works and of what a protocol is.

It should also be understood that GPS is not the model of Newell and Simon's theory. GPS implies more about a distinct level of generality independent of the tasks to be accomplished than their theory requires. Indeed, there does not exist any one computer program that is a model of their theory. Instead there exists a number of programs, by no means all of them composed by Newell and Simon or their co-workers, that are substantially consistent with the theory and that employ the "main methods of GPS" listed above. It is the information-processing theory of man which concerns us here, not GPS as such. And we are concerned with that theory precisely because it, in one variation or another, sometimes explicitly and sometimes implicitly, underlies almost all the new information-processing psychology and constitutes virtually a dogma for the artificial-intelligence community.

The basic conclusions the theory reaches are the following.

"All humans are information processing systems, hence have certain basic organizational features in common; all humans have in common a few universal structural characteristics, such as nearly identical memory parameters. These commonalities produce common characteristics of behavior among all human problem solvers.

"Since the information processing system [i.e., the human seen as an information-processing system—J. W.] can be factored into (1) basic structure, and (2) the contents of long-term memory [i.e., programs and data], it follows that any proposal for commonality among problem solvers not attributable directly to basic structure must be represented as an identity or similarity in the contents of the long-term memories—in the production system or in other stored memory structures."

[The theory] "proposes a system that, given enough time, can absorb any specification whatsoever—can become responsive to the full detail, say, of an encyclopedia (or a library of them). Hence the theory places the determination of differences and similarities of behavior directly upon the causes defining the content that will be stored in the human long-term memory. But these determinants of content are largely contingent upon the detail of the individual's life history. This does not mean that the determining processes are arbitrary or capricious or unlawful. It means that the contents can be as varied as the range of physical, biological, and social phenomena that surround the individual and from which he extracts them."[17] *

What is so remarkable about these conclusions is their scope, e.g., that the system the theory proposes, presumably a GPS-like system, can absorb any specification whatsoever. This claim—and what else can it reasonably be called?—is consistent with others of the authors' claims, namely, that the theory can account for the whole man and that computers will, within the visible future, handle problems over the whole range of human thought. The absurdity of what is being claimed for a GPS-like system is underscored by Newell and Simon's assertion that "The apparently complex behavior of the information processing system in a given environment is produced by the interaction of the demands of that environment with a few basic parameters of the system, particularly characteristics of its memories."[18] This is of course entirely consistent with their belief that man (like the ant) is "quite simple." But in this context a

---

* These statements invite comparison with B. F. Skinner's: "A scientific analysis of behavior must, I believe, assume that a person's behavior is controlled by his genetic and environmental histories rather than by the person himself as an initiating, creative agent" (from his *About Behaviorism*, New York, Alfred A. Knopf, New York, 1974, p. 189). The only difference between Skinner's position and that of the theory under discussion—and this difference is important from one point of view but totally irrelevant from another—is that Skinner refuses to look inside the black box that is the person, whereas the theory sees the inside as a computer.

technical claim is being made, namely, that GPS is quite simple, in the sense that, by changing a *few* of its parameters, its interaction with its environment will produce appropriately varied behavior simulating that of man.

In ordinary technical discussion we speak of a system being sensitive to "a few parameters" when the whole of its relevant mode of behavior can be entirely predetermined by setting a few switches or by entering a few data into its information store. A ship's navigation computer is of this type, for example. It will navigate the ship anywhere given only the geographical coordinates of its destination, some weather data, and so on. But to convert a GPS system from a chess player, say, to a cryptarithmetic puzzle solver is not a matter of changing a few numbers. In effect, the entire "memory structure" of GPS has to be replaced whenever GPS is to switch from one task to another. In other words, GPS is essentially nothing more than a programming language in which it is possible to write programs for certain highly specialized tasks. But, unless a computer program is to be considered a single parameter, GPS does not constitute any support for the claim that the complexity of human behavior is a function of only the human environment and a few parameters internal to the human information-processing system.

Occasionally, Newell and Simon do express a note of caution as when, for example, they admit that "we do not know what part of all human problem-solving activity employs a problem space, but over the range of tasks and individuals we have studied—a broad enough spectrum to make the commonalities nontrivial—a problem space is always used." But then, such mild disclaimers are countered by statements such as, "In spite of the restricted scope of the explicit evidential base of the theory, we will put it forth as a general theory of problem solving, without attempting to assess the boundaries of its applicability," and "we believe that the theory we are putting forth is much broader than the specific data on which we are erecting it."[19]

It is precisely this unwarranted claim to universality that demotes their use of the computer, computing systems, programs, etc., from the status of a scientific theory to that of a metaphor. They themselves say it: "Something ceases to be metaphor when detailed calculations can be made from it; it remains metaphor when it is rich in features in its own right, whose relevance to the object of comparison is problematic."[20] The question then is, can detailed calculations be made from their "theory"? (I shall continue to use the word "theory" here, since it would be too awkward to always write "alleged theory" when referring to the work in question.)

The answer seems, at first glance, to be a resounding "yes." Is not Newell and Simon's book filled with examples of calculations made by GPS? But there is a subtle point here, a point of great importance, a point almost universally overlooked by workers in artificial intelligence who also believe themselves to be in possession of genuine theories. This point is perhaps most clearly illuminated by contrasting Feigenbaum's rote-memory simulator with the GPS programs reported in Newell and Simon's book. Feigenbaum's program is, as I said earlier, a model of a psychological theory, that is, of how people struggle with the task of memorizing nonsense syllables. The program itself is also a theory, as I pointed out; for example, if it is given to a psychologist who is familiar with the programming language in which it is written, one may expect that he will understand it. The property it has which qualifies it as theory, however, is that it enunciates certain principles from which consequences may be drawn. These principles themselves are in computer-program form, and their consequences emerge in the behavior of the program, that is, in the computer's reading of the program. Among them are the well-known phenomena of interference and retroactive inhibition that I mentioned earlier.

The situation is entirely different when, say, the logic-theory program is run in GPS. To be sure, the LOGIC THEORIST is again a theory (albeit a quite trivial one), specifically, a theory of how novices go about solving certain elementary logic problems. But GPS, and this is the crucial point, *is merely a framework within which the logic-theory program runs*. GPS is, in effect, a programming language which it is relatively easy to write logic-theory programs, cryptarithmetic programs, and so on. The elementary information processes, the eip's, which constitute its elementary instructions are simply the primitive instructions of the machine into which GPS has transformed its host computer. GPS as such does not contain any principles—unless one counts as principles such observations as that, to solve problems, one must operate in terms of very general symbolic structures representing objects, operators, features of objects, and differences between objects, that one must build up a library of methods, and so on. Even then, GPS does not permit one to draw consequences from such "principles."

To say that GPS is, in any sense at all, an embodiment of a theory of human problem solving is equivalent to saying that high-school algebra is also such an embodiment. It too is a language, a computing schema, within which one can represent a theory already arrived at by other means. There is, of course, a theory of algebra. And there are theories of programming languages. But neither pretends to say anything about the psychology of human problem solving.

The counterargument to the above thesis is that the theory proposes a system—a GPS-like system—that embodies "a commonality among problem solvers" in its basic structure. It is that basic structure, that embodiment of the commonality among problem solvers, which makes it relatively easy to write problem-solving programs in a variety of quite disparate areas, e.g., logic and cryptarithmetic, in a GPS-like system. But, as Newell and Simon themselves said, any such commonality, if not attributable directly to basic structure, must be represented either in the program written in the GPS formalism or in the stored memory structures. In fact, all current versions of GPS-like systems have such absolutely minimal structure that the information that must be given them (in the form of program and data) for any particular problem-solving task must be detailed and specific, i.e., must define what the relevant operators are, to what objects they may be applied, what "difference" they make when applied to the proper object, and so on. As Newell and Simon say

"Due account must be taken of the limitations of GPS's access to the external world. The initial part of the explicit instructions to GPS have been acquired long ago by the human in building up his general vocabulary. This [information] has to be spelled out to GPS."[21]

There, precisely, is where the question is begged. For the real question is, what happens to the whole man as he builds his general vocabulary? How is his perception of what a "problem" is shaped by the experiences that are an integral part of his acquisition of his vocabulary? How do those experiences shape his perception of what "objects," "operators," "differences," "goals," etc., are relevant to any problem he may be facing? And so on. No theory that sidesteps such questions can possibly be a theory of human problem solving.

The dream of the artificial intelligentsia—a happy phrase the world owes to Dr. Louis Fein—is, of course, to bring into the world "machines that think, that learn, and that create," and whose ability to do these things will increase until "the range of problems they can handle will be coextensive with the range to which the human mind has been applied," as Drs. Newell and Simon already announced in 1958.[22] Their book was published fourteen years later and, as they promised, "the [machines'] ability to do these things increased rapidly," although the then "visible future" appears not to have arrived yet. But the vision is still clear enough. Now, indeed, they have told us how the trick is to be achieved. The proposed system, given enough time (but within the visible future), will become responsive to the full detail of a library of encyclopedias. In order for it to become thus responsive, however, it too will have to acquire a general vocabulary comparable to that commanded by an adult human; it will have to master natural language and internalize a fund of knowledge coextensive with that commanded by the human mind. A large segment of the artificial-intelligence community is, in fact, concentrating on the problem of computer understanding of natural language. That is the problem I intend to discuss in the next chapter.

For the moment, however, it remains to ask what image of man as problem solver can engender—I will not say justify—the mind-boggling vision here presented? To answer that question, we must look, first of all, at what Newell and Simon mean by a "problem."

Newell and Simon write,

"If we provide a representation for [what is desired, under what conditions, by means of what tools and operations, starting with

what initial information, and with what access to resources], and
assume that the interpretation of [the symbol structures that represent this information] is implicit in the program of the problem-
solving information-processing system, then we have defined a
problem."[23]

And then, of course, since "all humans are information-processing
systems," one can apply to them and to their affairs the "main methods of GPS," that is, "heuristic means-ends analysis," the testing of
objects to see if they are "undesired" and therefore yet to be transformed by operators into the "desired objects," and so on.

It may be objected that such a characterization of the aims of
artificial intelligence is a playing with words that unjustly overstates
AI's actual and much more modest goals, that Newell and Simon and
the AI community generally are really only talking about a certain
class of technical problems to which the above definition applies and
for which GPS-like methods are surely appropriate. But the point is
precisely that the pervasion—we might well say perversion—of ev-
eryday thought by the computer metaphor has turned every problem into a technical problem to which the methods here discussed
are thought to be appropriate. I shall have more to say on that theme
later.

Let it suffice for now to note that H. A. Simon had already
written in 1960,

"Let us suppose that a specific technological development permits the automation of psychiatry itself, so that one psychiatrist
can do the work formerly done by ten. . . . This example will seem
entirely fanciful only to persons not aware of some of the research
now going on into the possible automation of psychiatric processes."[24]

The research he had in mind was that then just begun by Kenneth
Mark Colby, a psychoanalyst, who wrote,

"Having conducted many laboratory experiments on free-associ-
ation and having had years of clinical experience with neurotic

processes, my initial hope was to simulate both [!] the free-associ-
ative thought characteristic of a neurotic process and its changes
under the influence of a psychotherapist's interventions."[25]

The project—happily—failed. But Simon's words were to ring in Dr.
Colby's ears for another six years before emerging again from his
own pen. As we have already noted, my own work on the ELIZA
system rekindled his enthusiasm and moved him to write the pas-
sages I quoted earlier but which bear repetition here:

> "If the [ELIZA] method proves beneficial, then it would provide
> a therapeutic tool which can be made widely available to mental
> hospitals and psychiatric centers suffering a shortage of thera-
> pists. . . . several hundred patients an hour could be handled by a
> computer system."[26]

Just as Simon predicted, and then some! Of course, this euphoric
promise is predicated precisely on a view of man as a GPS-like
machine. As Dr. Colby said,

> "A human therapist can be viewed as an information processor
> and decision maker with a set of decision rules which are closely
> linked to short-range and long-range goals. . . . He is guided in
> these decisions by rough empiric rules telling him what is appro-
> priate to say and not to say in certain contexts."[27]

The patient is, in other words, an object different from the desired
object. The therapist's task is to detect the difference, using differ-
ence-detecting operators, and then to reduce it, using difference-
reducing operators, and so on. That is his "problem"! And that is
how far the computer metaphor has brought some of us.

7

THE COMPUTER AND
NATURAL LANGUAGE

We distinguished three modes of artificial-intelligence research: the so-called performance, simulation, and theory modes. We observed, however, that the distinctions between them are not absolutely sharp. Moreover, we concluded that "theory" as used in the term "theory mode" has to be taken somewhat less than literally. The use of ideas derived from computers and computation in attempts to understand the human mind is rather more metaphorical than, say, is the use of mechanistic ideas in the understanding of the physical universe. But if we leave aside the vast body of work on modern computer science that deals either with theoretical issues concerning computation itself (e.g., finite automata theory or the theory of the structure of programming languages) or with the direct application of computers to specific tasks, independent of whether the execution of such tasks would count as intelligent behavior if it were accomplished by a human (e.g., the solving of systems of differential equations or the computer control of some complex chemical process), we are left with a subdomain of computer science in which at least one of the major aims is the imitation of man by machine. It will not prove useful for the purposes of this chapter to emphasize the various ways in which the work in this domain may be assigned primarily to psychology or to linguistics or to whatever established discipline. I shall therefore not press such distinctions in what follows.

Two things are clear: If we wish a machine to do something, we have to tell it to do it, and the machine must be able to understand what we say to it. The most common way to tell a computer what to do, at least to this day, is to give it a specific program for the task we have in mind and, of course, the data to which that program is to be applied. We may, for example, give it a square-root program and the number 25, and expect it to deliver the number 5 to us. The computer "understands" the square-root program in the sense that it can interpret it in precisely the way we had in mind when we composed it. But then such a program converts a computer into a very special-purpose machine, a square-root-taking machine, and nothing more. Humans, if they are machines at all, are vastly general-purpose machines and, what is most important, they understand communications couched in natural languages (e.g., English) that lack, by very far, the precision and unambiguousness of ordinary programming languages. Since the over-all aim of AI is to build machines that are "responsive to the full detail of a library of encyclopedias," work must naturally be done to enable them to understand natural language. But, even apart from such dreams, there are both practical and scientific reasons for working on the natural-language problem. If people from outside the computer fields are to be able to interact significantly with computers, then either they must learn the computer's languages or it must learn theirs. Even now it is easier to give computers the jargon-laden languages of some specialists—e.g., some physicians, or researchers working on moon-rocks—than it is to train the specialists in the ordinary languages of computers. Some computer scientists believe their theories about language to be somehow not fully legitimate as long as they remain what the general public disdainfully calls "mere theories," that is, until it has been shown that they can be converted into models in computer-program form. On the other hand, many linguists, for example, Noam Chomsky, believe that enough thinking about language remains to be done to occupy them usefully for yet a little while, and that any effort to convert their present theories into computer models would, if attempted by the people best qualified, be a diversion from the main task. And they rightly see no point to spending any of their energies studying the work of the hackers.

But to the truly initiated member of the artificial intelligentsia, no reason for working on the problem of machine understanding of natural language need be stated explicitly. Man's capacity to manipulate symbols, his very ability to think, is inextricably interwoven with his linguistic abilities. Any re-creation of man in the form of machine must therefore capture this most essential of his identifying characteristics.

There is, of course, no *single* problem that can reasonably be called the natural-language problem for computers, just as there is no such single problem for man. Instead, there are many problems, all having to do with enabling the computer to understand whatever messages are impressed on it from the world outside itself. The problem of computer vision, for example, is in many respects fundamentally the same as that of machine understanding of natural language. However the machine is made to derive information from its environment, it must, in some sense, "understand" it; that is, the computer must somehow be able to extract the semantic content from the messages that impinge on it, in part from their purely syntactic structure. It may seem odd, at first glance, to speak of the syntactic structure of a visual scene and to relate the process of understanding it to the process of understanding a natural-language text. But consider a picture of an adult and a child on a teetertotter. We understand certain aspects of that scene from its form, although even that understanding depends on our first having adopted a certain conceptual framework, a set of conventions. These conventions are syntactic in that they serve as criteria that permit us to distinguish legally admissable pictures, so to speak, from absurd ones. The ordinarily accepted picturing conventions would reject as ungrammatical most of the drawings of Escher, for example. We understand the teetertotter picture on the basis of semantic cues as well, however. We know, for example, that the adult figure, being down, is heavier than the child sitting high on the other side. And that knowledge comes to us from something other than the form of the picture, for it involves our private knowledge of aspects of the real world.

Language understanding, whether by man or machine, is like that too. We all have some criteria, an internalized grammar of the English language, that allow us to tell that the string of words "The house blue it" is ungrammatical. That is a purely syntactic judgment. On the other hand, we recognize that the sentence "The house blew it" is grammatical, even though we may have some difficulty deciding what it means, that is, how to understand it. We say we understand it only when we have been able to construct a story within which it makes sense, that is, when we can point to some contextual framework within which the sentence has a meaning, perhaps even an "obvious" meaning. For example, in a story in which a gambling house's scheme for beating a gambler's system misfired, the sentence "The house blew it" has a perfectly obvious meaning, at least to an American. Again, knowledge of the real world had to be brought to bear, not merely to disambiguate the sentence, but to assign meaning to it at all.

It is, of course, far easier to get a grip on the problem of machine understanding of natural language than on the corresponding problem for vision, first of all because language can be represented in written form, that is, as a string of symbols chosen from a very small alphabet. Moreover, such strings can be presented to the computer serially just as they are presented to human readers. They can also be stored with absolute fidelity. In contrast, the question of what constitutes a visual symbol, however primitive, already drags in major problems of both syntax and semantics. A worker on machine understanding of English text makes no important intellectual commitment to any particular research hypothesis or strategy when he adopts certain symbols as primitive, that is, as not themselves analyzable. But the worker on vision problems will have virtually determined major components of his research strategy the moment he decides on, say, edges and corners as elements of his primitive vocabulary. Besides, he faces a formidable problem just in deciding when his machine "sees" an edge. For this reason, as well as because it is only recently that television cameras have been coupled to computers, work on natural-language understanding by computers has a much longer history in artificial-intelligence research than does work on the problem of computer vision.

It has happened many times in the history of modern computation that some technological advance in computer hardware or programming (software) has triggered a virtually euphoric mania. When what were then thought of as large-scale computers first began to work more or less reliably, some otherwise reasonable people fell victim to what I will call the "Now that we have X (at last), we can do Y" syndrome. In this situation the X was what were then considered very large information stores (memories) and very high computing speeds, and the Y was machine translation of languages. (I shall not cite references here, the fever that plagued the afflicted having long ago subsided.)

The early vision was, as Robert K. Lindsay was to later put it "that high quality translations could be produced by machines supplied with sufficiently detailed syntactic rules, a large dictionary, and sufficient speed to examine the context of ambiguous words for a few words in each direction."[1] Computers are still not producing "high-quality translations." However, a hardcore dogmatist of the old school, if there is one left, might argue that we still don't have "sufficiently detailed syntactic rules" or "sufficient speed" to reach the desired end. But the real question is whether such sufficiency is possible at all. Would any set of syntactic rules, however detailed, and any computing speed, and any size dictionary suffice to produce high-quality translations? Every serious worker now agrees that the answer to this question is simply "no."

Translation must be seen as a process involving two distinct but not quite separable components: the text to be translated has to be understood; and the target-language text has to be produced. We can ignore the second of these components for our purposes here. The problem shows up in nearly its full complexity if we consider the target language to be the same as the source language and thus transform the translation problem into "simply" the paraphrasing problem. We have seen that to understand even a single sentence may involve both an elaborate contextual framework—e.g., a scenario having to do with gambling houses, gamblers' systems, and so on—and real-world knowledge—e.g., what gamblers do, what it means to break the bank, and so on. Let us return to that analogy. Suppose the sentence we cited, "The house blew it," occurred in the first chapter of a detective story. The detective's solution of the crime might hinge on his coming to understand that this sentence referred to gambling houses. But the clues that lead to that interpretation of the sentence may be revealed only gradually, say, one in each chapter. Then no man and no computer could be expected to understand, hence to paraphrase, that sentence when it first appears. Nor would an examination of a few words on either side of the sentence be any help whatever. Both a man and a computer would have to read all but the last chapter of the detective story to be able to do what the detective finally did, that is, understand the crucial sentence. (We assume that the story's last chapter serves only those who have missed a clue or have otherwise been unable to reach the appropriate conclusions.) And even then, only those with appropriate knowledge of the world could do it.

The recognition that a contextual framework is essential to understanding natural text was first exploited by so-called question-answering systems. B. F. Green and others wrote a system in 1961 that was able to understand and respond to questions about baseball, for example.[2] It could understand the question "Where did each team play in July?" without difficulty because, in its universe of discourse, such possibly problematic words as "team" and "play" could have only unique meanings. It could answer because each unambiguously understood question could easily be converted into a small program for searching the system's data base for relevant information. Bobrow's program **STUDENT**, although very much more ambitious, exploited the same principle.[3] It was able to solve so-called algebra word problems such as "Tom has twice as many fish as Mary has guppies. If Mary has 3 guppies, what is the number of fish Tom has?" Again, the universe of discourse within which the program was designed to operate determined how words and sentences were to be understood, reconstructed (into algebraic formulas), and otherwise manipulated. Note for example that, in order to "understand" the quoted problem, words like "fish" and "guppies" need not be "understood" at all; they could as well have been *"X"* and *"Y"*, respectively. And the word "has" has no connotation such as it would have in the sentence "Tom has a cold." The specification of a very highly constrained universe of discourse enormously simplifies the task of understanding—and that is, of course, true for human communication as well.

Obviously, understanding must be mutual in most realistic situations. In the context of man-machine communication, we wish the machine to understand us in order that it may do something for us, e.g., answer a question, solve a mathematical problem, or navigate a vehicle, which action we, in turn, hope to understand. The examples just cited shed no light on this aspect of man-machine communication. The answers delivered by either the **BASEBALL** or the **STUDENT** program simply do not have sufficient interpretive scope to be problematical. One cannot imagine having an interesting conversation with them. Among other things, and most significantly, they do not themselves ask questions.

The first program that illuminated this other side of the man-machine communication problem was my own **ELIZA**.[4]* **ELIZA** was a program consisting mainly of general methods for analyzing sentences and sentence fragments, locating so-called key words in texts, assembling sentences from fragments, and so on. It had, in other words, no built-in contextual framework or universe of discourse. This was supplied to it by a "script." In a sense **ELIZA** was an actress who commanded a set of techniques but who had nothing of her own to say. The script, in turn, was a set of rules which permitted the actor to improvise on whatever resources it provided.

The first extensive script I prepared for **ELIZA** was one that enabled it to parody the responses of a nondirective psychotherapist in an initial psychiatric interview. I chose this script because it enabled me to temporarily sidestep the problem of giving the program

---

* I chose the name "Eliza" because, like G. B. Shaw's Eliza Doolittle of *Pygmalion* fame, the program could be taught to "speak" increasingly well, although, also like Miss Doolittle, it was never quite clear whether or not it became smarter.

a data base of real-world knowledge. After all, I reasoned, a psychiatrist can reflect the patient's remark, "My mommy took my teddy bear away from me," by saying, "Tell me more about your parents," without really having to know anything about teddy bears, for example. In order to generate this response, the program had to know that "mommy" means "mother" and that the "patient" was telling it something about one of his parents. Indeed, it gleaned more than that from the subject's input, some of which, for example, it might use in later responses. Still, it could have been said to have "understood" anything in only the weakest possible sense.[5]

Nevertheless, ELIZA created the most remarkable illusion of having understood in the minds of the many people who conversed with it. People who knew very well that they were conversing with a machine soon forgot that fact, just as theatergoers, in the grip of suspended disbelief, soon forget that the action they are witnessing is not "real." This illusion was especially strong and most tenaciously clung to among people who knew little or nothing about computers. They would often demand to be permitted to converse with the system in private, and would, after conversing with it for a time, insist, in spite of my explanations, that the machine really understood them.

This phenomenon is comparable to the conviction many people have that fortune-tellers really do have some deep insight, that they do "know things," and so on. This belief is not a conclusion reached after a careful weighing of evidence. It is rather a hypothesis which, in the minds of those who hold it, is confirmed by the fortune-teller's pronouncements. As such, it serves the function of the drunkard's lamppost we discussed earlier: no light is permitted to be shed on any evidence that might be disconfirming; and, indeed, anything that might be seen as such evidence by a disinterested observer is interpreted in a way that elaborates and fortifies the hypothesis.

Within limits, this is a quite normal and even necessary process. No "information" is data except in the light of some hypothesis. Therefore, even in an ordinary two-person conversation, each participant brings something of himself to bear on the process of understanding the other. Each has, in other words, a working hypothesis, again a conceptual framework, concerning who the other is and what the conversation is about. This hypothesis serves as a predictor of what the other is going to say and, more importantly, of what he intends to mean by what he is going to say. This predictor functions simultaneously at several distinct levels. On the lowest level, a listener anticipates what the speaker's next few words will be; he completes yet unfinished sentences for him. If, for example, one is on an elevator and hears the operator say "This elevator does not stop on the . . . ," one would expect him to complete the sentence with the word "floor," but not with "second Thursday of each month." Sometimes the listener predicts wrongly and repairs the resulting damage only much later in the conversation, that is, when overwhelming evidence that he must have "misheard" is presented to him. Often, however, the erroneous prediction is falsified before the sentence in question has been completed by the speaker. The listener then makes corrections on the fly and virtually unconsciously.

On a much higher level, each participant brings to the conversation an image of who the other is. Since it is impossible for any human to know another completely, that image consists in part of attributions to the other's identity, attributions which must necessarily be based on evidence derived from independent life experiences of the participant. Our recognition of another person is thus an act of induction on evidence presented to us partly by him and partly by our reconstruction of the rest of the world; it is a kind of generalization. We are, in other words, all of us prejudiced—in the sense of pre-judging—about each other. And, as we have noted, we all find it hard, or even nearly impossible, to perceive—let alone to accept and to permit to become operative—evidence that tends to disconfirm our judgments.

It is then easy to understand why people conversing with ELIZA believe, and cling to the belief, that they are being understood. The "sense" and the continuity the person conversing with ELIZA perceives is supplied largely by the person himself. He assigns meanings and interpretations to what ELIZA "says" that confirm his initial hypothesis that the system does understand, just as he might do with what a fortune-teller says to him. All ELIZA or the fortune-teller need do is give responses that are sufficiently plausible and that allow a sufficient scope for interpretation to make such constructions possible. And, since the subject cannot probe the true limits of ELIZA's capacities (he has, after all, only a limited time to play with it, and it is constantly getting new material from him), he cannot help but attribute more power to it than it actually has. Besides, he knows that ELIZA was constructed by a professor at a university. It is therefore clothed in the magical mantle of Science and all of Science's well-known powers may be attributed to it.

ELIZA did, in fact, generate plausible responses to what was said to it. In order to be able to do that, it too had to be supplied with a set of expectations. These were encoded in whatever script was given to it. A person playing with ELIZA in its psychiatrist mode was instructed to provide ELIZA with the sort of statements one might make to a psychiatrist in an initial psychiatric interview. He was told, in other words, what ELIZA's expectations were. On a lower level, ELIZA's psychiatric script was constructed in a way that allowed ELIZA to make local predictions about sentences and textual fragments, that is, to apply hypotheses to them which further examinations might confirm or falsify. For example, the psychiatric script entertained the initial hypothesis that a fragment of the general form "everybody . . .me," although patently conveying a message about the subject's relationship to "everybody," e.g., "everybody hates me," or about what everybody is doing to the subject, e.g., "everybody is always laughing at me," latently and more importantly referred to a recent incident involving the subject and only a single or at most a few individuals. ELIZA's response might therefore be "Tell me, who told you he hated you within the last few days?" or "Who laughed at you recently?"

What sharply distinguishes the current work on machine understanding of natural language from the work of the early 1960's and before is precisely the current strong use of prediction, both on the local syntactic level and, more importantly, on the larger contextual level. Roger C. Schank, an exceptionally brilliant young representative of the modern school, bases his theory on the central idea that every natural-language utterance is a manifestation, an encoding, of an underlying conceptual structure. Understanding an utterance means encoding it (Schank uses the technical term "mapping") into one's own internal conceptual structure.

> "Any two utterances that can be said to mean the same thing, whether they are in the same or different languages, should be characterized in only one way by the conceptual structures. . . . The representation of this conceptual content then, must be in terms that are interlingual and as neutral as possible. . . . We will be . . . [concerned] with finding, once something is said, a representation that will account for the meaning of that utterance in an unambiguous way and one that can be transformed back into that utterance or back into any other utterances that have the same meaning.
>
> "The important point is that underlying every sentence in a language, there exists at least one conceptualization."[6]

What I wish to emphasize here is that Schank's theory proposes a formal structure for the conceptual bases underlying linguistic utterances, that it proposes specific mechanisms (algorithms) for basing predictions on such conceptual structures, and that it proposes formal rules for analyzing natural-language utterances and for converting them into the conceptual bases. However, Schank does not believe that an individual's entire base of conceptions can be explicitly extricated from him. He believes only that there exists such a belief structure within each of us, and that, if it could be explicated, it could in principle be represented by his formalism. One difficulty, which Schank of course recognizes, is that every individual's belief structure is constantly changing.

In discussing the role a person's belief structure plays in the way he participates in conversations, I wrote in my 1967 "Contextual Understanding" paper

> "In some areas of the individual's intellectual life, this structure may be highly logically organized—at least up to a point; for example, in the area of his own profession. In more emotionally loaded areas, the structure may be very loosely organized and even contain many contradictions. When a person enters a conversation he brings his belief structures with him as a kind of agenda.
>
> "A person's belief structure is a product of his entire life experience. All people have some common formative experiences, e.g., they were all born of mothers. There is consequently some basis of

understanding between any two humans simply because they are human. But even humans living in the same culture will have difficulty in understanding one another where their respective lives differed radically. Since, in the last analysis, each of our lives is unique, there is a limit to what we can bring another person to understand. There is an ultimate privacy about each of us that absolutely precludes full communication of any of our ideas to the universe outside ourselves and which thus isolates each one of us from every other noetic object in the world.

"There can be no total understanding and no absolutely reliable test of understanding.

"To know with certainty that a person understood what has been said to him is to perceive his entire belief structure and *that* is equivalent to sharing his entire life experience. It is precisely barriers of this kind that artists, especially poets, struggle against.

"This issue must be confronted if there is to be any agreement as to what machine "understanding" might mean. What the above argument is intended to make clear is that it is too much to insist that a machine understands a sentence (or a symphony or a poem) only if that sentence invokes the same imagery in the machine as was present in the speaker of the sentence at the time he uttered it. For by that criterion no human understands any other human. Yet, we agree that humans do understand one another to *within acceptable tolerances*. The operative word is "acceptable" for it implies *purpose*. When, therefore, we speak of a machine understanding, we must mean understanding as limited by some objective. He who asserts that there are certain ideas no machines will ever understand can mean at most that the machine will not understand these ideas tolerably well because they relate to objectives that are, in his judgment, inappropriate with respect to machines. Of course, the machine can still deal with such ideas symbolically, i.e., in ways which are reflections—however pale—of the ways organisms for which such objectives are appropriate deal with them."[7]

I would expect Schank, as well as most other workers now in this field, to find this consistent with their own ideas. However, when I used the term "imagery" ("that sentence invokes the same imagery"), a term I now see as roughly corresponding to Schank's "conceptual structures," I had no idea at all about how such images might be represented in a formal system.

There exist today several computer language-understanding systems that rely on ideas that resemble Schank's much more than superficially, even though they were arrived at independently and do differ in important respects. One of the best of these, and also one of the better known, is that of Terry Winograd. Winograd, at the time a graduate student in M.I.T.'s Artificial Intelligence Laboratory, was working in close association with a group building a computer-controlled "hand-eye" machine, that is, a computer system that could "see" its environment by means of an attached television camera, and that could manipulate objects within its environment by means of a computer-controlled mechanical arm and hand. One problem this system solved—to give an example of its general capability—was to copy a structure made of children's blocks, e.g., one consisting of arches, bridges, and so on, on the basis of a model of the desired structure that had been shown to it. Winograd addressed himself to the task of developing a system to enable humans to instruct this robot, strictly with respect to the toy world of blocks within which it was to operate, in natural language.

> "Within this mini-world, we can give the computer a deep kind of knowledge. . . . [the robot] can manipulate toy blocks on a table containing simple objects like a box. In the course of a dialogue, it can be asked to manipulate the objects, doing such things as building stacks and putting things into the box. It can be questioned about the current configurations of blocks on the table, about the events that have gone on during the discussion, and to a limited extent about its reasoning. It can be told simple facts which are added to its store of knowledge for use in later reasoning. The conversation goes on within a dynamic framework—one in which the computer is an active participant, doing things to change his toy world, and discussing them."[8]

Following is a sample conversation with the robot.

Human: Find a block which is taller than the one you are holding and put it into the box.

Robot: By "it," I assume you mean the block which is taller than the one I am holding.

H: OK.

R: (The hand sets down the block it is holding and picks up a taller one.)

H: What does the box contain?

R: The blue pyramid and the blue box.

H: What is the pyramid supported by?

R: The box.

. . .

H: Why did you clear off [the red] cube?

R: To put it on a large green cube.

H: Why did you do that?

R: To stack up the red cube, a large red block, and a large green cube.

H: Why did you do that?

R: Because you asked me to.[9]

Winograd's program is an important achievement. It shows how a specific view of certain aspects of language can really be filled in with enough detail to provide a working model. It consists fundamentally of procedural descriptions, i.e., programs, of cognitive processes and is thus an embodiment of a particular psychological approach to language understanding as a process. On the other hand, the program has serious shortcomings, some of which Winograd himself notes:

[It does not deal] "with all the implications of viewing language as a process of communication between two intelligent people. A human language user is always engaged in a process of trying to understand the world around him, including the person he is talking to. He is actively constructing models and hypotheses, and he makes use of them in the process of language understanding."

[Because the system] "keeps track of when things have been mentioned, it can check a possible interpretation of a question to see whether the asker could answer it himself from his previous sentences. If so, it assumes he probably means something else. We could characterize this as containing two sorts of knowledge. First, it assumes that a person asks questions for the purpose of getting information he doesn't already have, and second, it has a very primitive model of what information he has on the basis of what he has said. A realistic view of language must have a complex model of this type, and the heuristics in our system touch only the tiniest bit of the relevant knowledge."[10]

It really must be said that this expression of humility is enormously refreshing, especially since it comes from within the priesthood of the artificial intelligentsia, and so is virtually unique. Unfortunately, it does not go far enough. For what Winograd has done—indeed, what all of artificial intelligence has so far done—is to build a machine that performs certain specific tasks, just as, say, seventeenth-century artisans built machines that kept time, fired iron balls over considerable distances, and so forth. Those artisans would have been grievously mistaken had they let their successes lead them to the conclusion that they had begun to approach a general theoretical understanding of the universe, or even to the conclusion that, because their machines worked, they had validated the idea that the laws of the universe are formalizable in mathematical terms. The *hubris* of the artificial intelligentsia is manifested precisely by its constant advance of exactly these mistaken ideas about the machines it has succeeded in building. Neither Winograd's humility, nor that of any other AI researcher, extends to that admission.

Newell, Simon, Schank, and Winograd simply mistake the nature of the problems they believe themselves to be "solving." As if they were benighted artisans of the seventeenth century, they present "general theories" that are really only virtually empty heuristic slogans, and then claim to have verified these "theories" by constructing models that do perform some tasks, but in a way that fails to give insight into general principles. The failure is intrinsic, for they have failed to recognize that, in order to do what they claim to do, they must discover and formulate general principles of more power than that inherent in the observation, or even the demonstration, that laws can be stated in the form of computer programs. The most important and far-reaching effect of this failure is that researchers in artificial intelligence constantly delude themselves into believing that the reason any particular system has not come close to realizing AI's grand vision is always to be found in the limitations of the specific system's program. Thus, for example, Winograd acknowledges that his system avails itself of only the tiniest bit of relevant knowledge. The knowledge he is talking about is the knowledge of "facts" that is available to humans. But the problem with his approach is that his heuristics express no interesting general principles. Furthermore, such principles cannot be discovered merely by expanding the range of a system in a way that enables it to get more knowledge of the world. Even the most clever clock builder of the seventeenth century would never have discovered Newton's laws simply by building ever fancier and more intricate clocks!

Artificial intelligence has, as we have documented, set as its goal the building of machines whose range of thought is to be coextensive with that of humanity itself. (Never mind, for now, whether this is to be achieved in the "visible future" or not.) And the theories that are to underpin this triumph of AI are to apply to the whole man as well. Clearly, then, the kinds of limitations to which presently existing systems are subject, and to some of which Winograd confesses with genuine humility, are seen by the AI community as a whole as being merely temporary difficulties that can be overcome— in the visible future, according to Newell and Simon. There are then, two questions that must ultimately be confronted. First, are the conceptual bases that underlie linguistic understanding entirely formalizable, even in principle, as Schank suggests and as most workers in AI believe? Second, are there ideas that, as I suggested, "no machines will ever understand because they relate to objectives that are inappropriate for machines"?

These two questions are of enormous importance. They go to the heart of the question about whether there is any essential difference between man and machine. And it is appropriate that they be asked in the context of a discussion of the problem of natural-language understanding by machines, for it is in his language, above all, that man manifests his intelligence and, some believe, his unique identity as man. The two questions also need to be asked together. They are inextricably linked to one another. For if the whole of a human experience and the belief structure to which it gives rise cannot be formalized, then there are indeed appropriately human objectives that are inappropriate for machines. And if we were to conclude (as I intend to) that there are indeed such objectives, then we could also say something about what machines ought and ought not to be put to doing.

The fact that these questions have become important at all is indicative of the depth to which the information-processing metaphor has penetrated both the academic and the popular mind. For when we take stock, we quickly discover how little has actually been accomplished so far. Newell and Simon's book, *Human Problem Solving,* speaks in detail of only three problems: cryptarithmetic, theorem proving in the simplest logical calculus, and chess. The accomplishments of the computer-controlled "hand-eye" machines at MIT and Stanford University (e.g., building block structures from models shown to them, and screwing nuts onto bolts), are rightly hailed as triumphs by those who understand the incredible complexity of the problems that had first to be solved. And there have been other triumphs of similar magnitude. But the very fact that such achievements deserve to be so applauded itself testifies to how utterly primitive is our current knowledge about the human mind. George A. Miller, like the modern computer linguists, also speaks of the conceptual structures that underlie human thought and language, but he says, "To pretend that we know how to impart these complex conceptual structures to any machine at the present time is simply absurd."[11] Winograd is really speaking for the entire field of artificial intelligence, whether his colleagues acknowledge that fact or not, when he says that our systems have touched only the tiniest bit of the relevant knowledge.

But just as it would have been unfair to argue in the seventeenth century that to place a manmade object into Earth orbit is impossible on the grounds that no one at that time had the slightest idea about how to accomplish it, so it would be wrong today to make impossibility arguments about what computers can do entirely on the grounds of our present ignorance. It is relevant, however, especially for taking stock of our present situation, to examine the power of the theories we have been discussing. Are they, for example, Newtonian in the vastness of their inferential scope?

What is contributed when it is asserted that "there exists a conceptual base that is interlingual, onto which linguistic structures in a given language map during the understanding process and out of which such structures are created during generation [of linguistic utterances]"?[12] Nothing at all. For the term "conceptual base" could perfectly well be replaced by the word "something." And who could argue with that so-transformed statement? Schank's contribution, like those of others now tilling the same fertile fields, is that he attempts to provide a formal representation of the conceptual base. He intends to tell us in utmost detail what that something "that underlies all natural languages" is and how it functions in both the generation and the understanding of linguistic utterances. Even then, Schank provides no demonstration that his scheme is more than a collection of heuristics that happen to work on specific classes of examples. The crucial scientific problem would be to construct a finite program that assigns appropriate conceptual structures to the infinite range of sentences that can occur in natural language. That problem remains as untouched as ever. Imagine an adding machine that adds some but not all numbers correctly, and about which we can't even say what characterizes the numbers it can add. We would hardly call that a mechanization of arithmetic. And in what form are the conceptual structures Schank hypothesizes, and the operations on them, linkages among them, and so on, to be shown to us? In the form of computer programs, of course.

> "What is needed, and what has been lacking, is a cohesive theory of how humans understand natural language without regard to particular subparts of that problem, but with regard to that problem as a whole. The theory . . . is also intended to be a basis for computer programs that understand natural language. . . . What will be discussed is the theory of such a program. . . .
>
> "We hope to be able to build a program that can learn, as a child does, how to do what we have described in this paper instead of being spoon-fed the tremendous information necessary. In order to do this it might be necessary to await an effective automatic hand-eye system and an image processor."[13]

Here we begin to see the confluence of the work on problem solving we discussed earlier and the work on natural-language understanding by machine. A sentence is a "given object," the conceptual structure that is its meaning is the "desired object," the goal is to transform the former into the latter, and the means are those provided by the understanding program. But what is most important in both instances is that the theories be convertible to computer programs.

It may be possible, following Schank's procedures, to construct a conceptual structure that corresponds to the meaning of the sentence, "Will you come to dinner with me this evening?" But it is hard to see—and I know that this is not an impossibility argument—how Schank-like schemes could possibly understand that same sentence to mean a shy young man's desperate longing for love. Even if a computer could simulate feelings of desperation and of love, is the computer then capable of *being* desperate and of loving? Can the computer then understand desperation and love? To the extent that those are legitimate questions at all, and that is a very limited extent indeed, the answer is "no." And if that is the answer, then the sense in which even the most powerful Schank-like system "understands" is about as weak as the sense in which ELIZA "understood."

At best, what we see here is another example of the drunkard's search. A theory purports to describe the conceptual structures that underlie all human language understanding. But the only conceptual structures it admits as legitimate are those that can be represented in the form of computer-manipulatable data structures. These are then simply pronounced to constitute all the conceptual structures that underlie all of human thought. Given such a program, i.e., such a narrowing of the meaning of the word "all," it should indeed be possible to prove that the theory accounts for "all" human linguistic behavior!

A theory is, of course, itself a conceptual framework. And so it determines what is and what is not to count as fact. The theories— or, perhaps better said, the root metaphors—that have hypnotized the artificial intelligentsia, and large segments of the general public as well, have long ago determined that life is what is computable and only that. As Professor John McCarthy, head of Stanford University's Artificial Intelligence Laboratory said, "The only reason we have not yet succeeded in formalizing every aspect of the real world is that we have been lacking a sufficiently powerful logical calculus. I am currently working on that problem."

Sometimes when my children were still little, my wife and I would stand over them as they lay sleeping in their beds. We spoke to each other in silence, rehearsing a scene as old as mankind itself. It is as Ionesco told his journal: "Not everything is unsayable in words, only the living truth."

8

ARTIFICIAL INTELLIGENCE

When Roger Schank expressed the hope that we will be able to build a program that can learn as a child does, he was echoing words spoken by H. A. Simon over ten years earlier:

> "If GPS is a theory of how a machine can bootstrap itself into higher intelligence or how people learn language, then let it bootstrap itself, and let it learn language. This is an entirely appropriate obligation to impose. . . . Not just on behalf of myself, but on behalf of the entire group of people working in the field, I accept the obligation and hope that one of us will produce the requisite programs before too long."[1]

Both Simon and Schank have thus given expression to the deepest and most grandiose fantasy that motivates work on artificial intelligence, which is nothing less than to build a machine on the model of man, a robot that is to have its childhood, to learn language as a child does, to gain its knowledge of the world by sensing the world through its own organs, and ultimately to contemplate the whole domain of human thought. (It is worth noting, though only by the way for now, that should this dream be realized, we will have a language-understanding machine but still no theory of language understanding as such, for observing a machine "learning as a child does" does not in itself constitute an understanding of the language-acquisition process.)

Whether or not this program can be realized depends on whether man really is merely a species of the genus "information-processing system" or whether he is more than that. I shall argue that an entirely too simplistic notion of intelligence has dominated both popular and scientific thought, and that this notion is, in part, responsible for permitting artificial intelligence's perverse grand fantasy to grow. I shall argue that an organism is defined, in large part, by the problems it faces. Man faces problems no machine could possibly be made to face. Man is not a machine. I shall argue that, although man most certainly processes information, he does not necessarily process it in the way computers do. Computers and men are not species of the same genus.

Few "scientific" concepts have so thoroughly muddled the thinking of both scientists and the general public as that of the "intelligence quotient" or "I.Q." The idea that intelligence can be quantitatively measured along a simple linear scale has caused untold harm to our society in general, and to education in particular. It has spawned, for example, the huge educational-testing movement in the United States, which strongly influences the courses of the academic careers of millions of students and thus the degrees of certification they may attain. It virtually determines what "success" people may achieve in later life because, in the United States at least, opportunities to "succeed" are, by and large, open only to those who have the proper credentials, that is, university degrees, professional diplomas, and so on.

When modern educators argue that intelligence tests measure a subject's ability to do well in school, they mean little more than that these tests "predict" a subject's ability to pass academic-type tests. This latter ability leads, of course, to certification and then to "success." Consequently, any correlation between the results of such tests and people's "success," as that term is understood in the society at large, must necessarily be an artifact of the testing procedure. The test itself has become a criterion for that with which it is to be correlated! "Psychologists should be ashamed of themselves for promoting a view of general intelligence that has engendered such a testing program."[2]

My concern here is that the mythology that surrounds I.Q. testing has led to the widely accepted and profoundly misleading conviction that intelligence is somehow a permanent, unalterable, and culturally independent attribute of individuals (somewhat like, say, the color of their eyes), and moreover that it may even be genetically transmittable from generation to generation.

The trouble with I.Q. testing is not that it is entirely spurious, but that it is incomplete. It measures certain intellectual abilities that large, politically dominant segments of western European societies have elevated to the very stuff of human worth and hence to the *sine qua non* of success. It is incomplete in two ways: first, in that it fails to take into account that human creativity depends not only on intellect but also crucially on an interplay between intellect and other modalities of thought, such as intuition and wisdom; second, in that it characterizes intelligence as a linearly measurable phenomenon that exists independent of any frame of reference.

Einstein taught us that the idea of motion is meaningless in and of itself, that we can sensibly speak only of an object's motion relative to some frame of reference, not of any *absolute* motion of an object. When, in speaking informally, we say that a train moved, we mean that it moved relative to some fixed point on the earth. We need not emphasize this in ordinary conversation, because the earth (or our body) is to us a kind of "default" frame of reference that is implicitly assumed and understood in most informal conversation. But a physicist speaking as a physicist cannot be so sloppy. His equations of motion must contain terms specifying the coordinate system with respect to which the motion they describe takes place.

So it is with intelligence too. Intelligence is a meaningless concept in and of itself. It requires a frame of reference, a specification of a domain of thought and action, in order to make it meaningful. The reason this necessity does not strike us when we speak of intelligence in ordinary conversation is that the required frame of reference—that is, our own cultural and social setting with its characteristic domains of thought and action—is so much with us that we implicitly assume it to be understood. But our culture and our social milieu are in fact neither universal nor absolute. It therefore behooves us, whenever we use the term "intelligence" as scientists or educators, to make explicit the domain of thought and action which renders the term intelligible.

Our own daily lives abundantly demonstrate that intelligence manifests itself only relative to specific social and cultural contexts. The most unschooled mother who cannot compose a single grammatically correct paragraph in her native language—as, indeed, many academics cannot do in theirs—constantly makes highly refined and intelligent judgments about her family. Eminent scholars confess that they don't have the kind of intelligence required to do high-school algebra. The acknowledged genius is sometimes stupid in managing his private life. Computers perform prodigious "intellectual feats," such as beating champion checker players at their own game and solving huge systems of equations, but cannot change a baby's diaper. How are these intelligences to be compared to one another? They cannot be compared.

Yet forms of the idea that intelligence is measurable along an absolute scale, hence that intelligences are comparable, have deeply penetrated current thought. This idea is responsible, at least in part, for many sterile debates about whether it is possible "in principle" to build computers more intelligent than man. Even as moderate and reasonable a psychologist as George A. Miller occasionally slips up, as when he says, "I am very optimistic about the eventual outcome of the work on machine solution of intellectual problems. Within our lifetime machines may surpass us in general intelligence."[3]

The identification of intelligence with I.Q. has severely distorted the primarily mathematical question of what computers can and cannot do into the nonsensical question of "how much" intelligence one can, again "in principle," give to a computer. And, of course, the reckless anthropomorphization of the computer now so common, especially among the artificial intelligentsia, couples easily to such simpleminded views of intelligence. This joining of an illicit metaphor to an ill-thought-out idea then breeds, and is perceived to legitimate, such perverse propositions as that, for example, a computer can be programmed to become an effective psychotherapist.

I had once hoped that it would be possible to prove that there is a limit, an upper bound, on the intelligence machines could achieve, just as Claude Shannon, the founder of modern information theory, proved that there is an upper bound on the amount of information a given information channel can transmit. Shannon proved that, for example, a specific telephone cable can carry at most a certain number of telephone conversations at any one time. However, before he could even sensibly formulate his now justly famous result, he had to have some way to quantify information. Else how could he speak of a channel's capacity to handle this "much" information but no "more"? Indeed, his design of an information measure itself constitutes an important contribution to modern science. (Given, of course, that he also founded a cogent theory within which his measure plays a decisive role.) It is now clear to me that, since we can speak of intelligence only in specific domains of thought and action, and since these domains are themselves not measurable, we can have no Shannon-like measure of intelligence and therefore no theorem of the kind I had hoped for. In plain words: we may express the wish, even the opinion, that there is a limit to the intelligence machines can attain, but we have no way of giving it precise meaning and certainly no way of proving it.

Does our inability to compute an upper bound on machine intelligence provide grounds either for the "optimistic" conclusion that "machines may surpass us in general intelligence" or for the very same "pessimistic" conclusion?* Neither. We learn instead that any argument that calls for such a conclusion, or for its denial, is itself ill-framed and therefore sterile.

These considerations shed additional light on a question alluded to in Chapter VII (p. 193), where I spoke of "objectives that are inappropriate for machines." Many people would argue that it is not

---

* The optimist says, "This is the best of all possible worlds!" The pessimist answers, "That's right."

reasonable to speak of machines as having objectives in the first place. But such a rhetorical quibble, if taken seriously, only begs the question, for it ignores the fact that people do in fact delegate responsibility to computers and give them objectives and purposes.

The question I am trying to pursue here is, "What human objectives and purposes may not be appropriately delegated to computers?" We can design an automatic pilot, and delegate to it the task of keeping an airplane flying on a predetermined course. That seems an appropriate thing for machines to do. It is also technically feasible to build a computer system that will interview patients applying for help at a psychiatric out-patient clinic and produce their psychiatric profiles complete with charts, graphs, and natural-language commentary. The question is not whether such a thing *can* be done, but whether it is appropriate to delegate this hitherto human function to a machine.

The artificial intelligentsia argue, as we have seen, that there is no domain of human thought over which machines cannot range. They take for granted that machines can think the sorts of thoughts a psychiatrist thinks when engaged with his patient. They argue that efficiency and cost considerations dictate that machines ought to be delegated such responsibilities. As Professor John McCarthy once put it to me during a debate, "What do judges know that we cannot tell a computer?" His answer to the question—which is really just our question again, only in different form—is, of course, "Nothing." And it is, as he then argued, perfectly appropriate for artificial intelligence to strive to build machines for making judicial decisions.

The proposition that judges and psychiatrists know nothing that we cannot tell computers follows from the much more general proposition subscribed to by the artificial intelligentsia, namely, that there is nothing at all which humans know that cannot, at least in principle, be somehow made accessible to computers.

Not all computer scientists are still so naive as to believe, as they were once charged with believing, that knowledge consists of merely some organization of "facts." The various language-understanding and vision programs, for example, store some of their knowledge in the form of assertions, i.e., axioms and theorems, and other of it in the form of processes. Indeed, in the course of planning and executing some of their complex procedures, these programs compose subprograms, that is, generate new processes, that were not explicitly supplied by human programmers. Some existing computer systems, particularly the so-called hand-eye machines, gain knowledge by directly sensing their environments. Such machines thus come to know things not only by being told them explicitly, but also by discovering them while interacting with the world. Finally, it is possible to instruct computers in certain skills, for example, how to balance a broomstick on one of its ends, by showing them how to do these things even when the instructor is himself quite incapable of verbalizing how he does the trick. The fact, then, and it *is* a fact, that humans know things which they cannot communicate in the form of spoken or written language is not by itself sufficient to establish that there is some knowledge computers cannot acquire at all.

But lest my "admission" that computers have the power to acquire knowledge in many diverse ways be taken to mean more than I intend it to mean, let me make my position very clear:

First (and least important), the ability of even the most advanced of currently existing computer systems to acquire information by means other than what Schank called "being spoon-fed" is still extremely limited. The power of existing heuristic methods for extracting knowledge even from natural-language texts directly "spoonfed" to computers rests precariously on, in Winograd's words, "the tiniest bit of relevant knowledge." It is simply absurd to believe that any currently existing computer system can come to know in any way whatever what, say, a two-year-old child knows about children's blocks.

Second, it is not obvious that all human knowledge is encodable in "information structures," however complex. A human may know, for example, just what kind of emotional impact touching another person's hand will have both on the other person and on himself. The acquisition of that knowledge is certainly not a function of the brain alone; it cannot be simply a process in which an information structure from some source in the world is transmitted to some destination in the brain. The knowledge involved is in part kinesthetic; its acquisition involves having a hand, to say the very least. There are, in other words, some things humans know by virtue of having a human body. No organism that does not have a human body can know these things in the same way humans know them. Every symbolic representation of them must lose some information that is essential for some human purposes.

Third, and the hand-touching example will do here too, there are some things people come to know only as a consequence of having been treated as human beings by other human beings. I shall say more about this in a moment.

Fourth, and finally, even the kinds of knowledge that appear superficially to be communicable from one human being to another in language alone are in fact not altogether so communicable. Claude Shannon showed that, even in abstract information theory, the "information content" of a message is not a function of the message alone but depends crucially on the state of knowledge, on the expectations, of the receiver. The message "Am arriving on 7 o'clock plane, love, Bill" has a different information content for Bill's wife, who knew he was coming home, but not on precisely what airplane, than for a girl who wasn't expecting Bill at all and who is surprised by his declaration of love.

Human language in actual use is infinitely more problematical than those aspects of it that are amenable to treatment by information theory, of course. But even the example I have cited illustrates that language involves the histories of those using it, hence the history of society, indeed, of all humanity generally. And language in human use is not merely functional in the way that computer languages are functional. It does not identify things and words only with immediate goals to be achieved or with objects to be transformed. The human use of language manifests human memory. And that is a quite different thing than the store of the computer, which has been anthropomorphized into "memory." The former gives rise to hopes and fears, for example. It is hard to see what it could mean to say that a computer hopes.

These considerations touch not only on certain technical limitations of computers, but also on the central question of what it means to be a human being and what it means to be a computer.

I accept the idea that a modern computer system is sufficiently complex and autonomous to warrant our talking about it as an organism. Given that it can both sense and affect its environment, I even grant that it can, in an extremely limited sense, be "social-ized," that is, modified by its experiences with its world. I grant also that a suitably constructed robot can be made to develop a sense of itself, that it can, for example, learn to distinguish between parts of itself and objects outside of itself, that it can be made to assign a higher priority to guarding its own parts against physical damage than to similarly guarding objects external to itself, and that it can form a model of itself which could, in some sense, be considered a kind of self-consciousness. When I say therefore that I am willing to regard such a robot as an "organism," I declare my willingness to consider it a kind of animal. And I have already agreed that I see no way to put a bound on the degree of intelligence such an organism could, at least in principle, attain.

I make these stipulations, as the lawyers would call them, not because I believe that what any reasonable observer would call a socialized robot is going to be developed in the "visible future"—I do not believe that—but to avoid the unnecessary, interminable, and ultimately sterile exercise of making a catalogue of what computers will and will not be able to do, either here and now or ever. That exercise would deflect us from the primary question, namely, whether there are objectives that are not appropriately assignable to machines.

If both machines and humans are socializable, then we must ask in what way the socialization of the human must necessarily be different from that of the machine. The answer is, of course, so obvious that it makes the very asking of the question appear ludi-crous, if indeed not obscene. It is a sign of the madness of our time that this issue has to be addressed at all.

Every organism is socialized by the process of dealing with problems that confront it. The very biological properties that differ-entiate one species from another also determine that each species will confront problems different from those faced by any other. Ev-ery species will, if only for that reason, be socialized differently. The human infant, as many observers have remarked, is born prema-turely, that is, in a state of utter helplessness. Yet the infant has biological needs which, if he is to survive at all, must be satisfied by others. Indeed, many studies of orphanages have shown that more than his merely elementary physical needs must be satisfied; an infant will die if he is fed and cleaned but not, from the very beginning of his life, fondled and caressed—if, in other words, he is not treated as a human being by other human beings.[4]

A catastrophe, to use Erik Erikson's expression for it, that every human being must experience is his personal recapitulation of the biblical story of paradise. For a time the infant demands and is granted gratification of his every need, but is asked for nothing in return. Then, often after the infant has developed teeth and has bitten the breast that has fed him, the unity between him and his mother is broken. Erikson believes this universal human drama to be the ontogenetic contribution to the biblical saga of the Garden of Eden. So important is this period in the child's life that

> "a drastic loss of accustomed mother love without proper substitution at this time can lead [under otherwise aggravating conditions] to acute infantile depression or to a mild but chronic state of mourning which may give a depressive undertone to the whole remainder of life. But even under the most favorable circumstances, this stage leaves a residue of a primary sense of evil and doom and of a universal nostalgia for a lost paradise."
>
> [These early stages] "then, form in the infant the springs of the basic sense of trust and the basic sense of mistrust which remain the autogenic source of both primal hope and of doom throughout life."[5]

Thus begins the individual human's imaginative reconstruction of the world. And this world, as I said earlier, is the repository of his subjectivity, the stimulator of his consciousness, and ultimately the constructor of the apparently external forces he is to confront all his life.

> "As the child's radius of awareness, co-ordination, and responsiveness expands, he meets the educative patterns of his culture, and thus learns the basic modalities of human existence, each in personally and culturally significant ways. . . . *To get* . . . means to receive and to accept what is given. This is the first social modality

learned in life; and it sounds simpler than it is. For the groping and
unstable newborn organism learns this modality only as it learns to
regulate its organ systems in accordance with the way in which the
maternal environment integrates its methods of child care. . . .

"The optimum total situation implied in the baby's readiness to
get what is given is his mutual regulation with a mother who will
permit him to develop and coordinate his means of getting as she
develops and co-ordinates her means of giving. . . . The mouth and
the nipple seem to be the mere centers of a general aura of warmth
and mutuality which are enjoyed and responded to with relaxation
not only by these focal organs, but by both total organisms. The
mutuality of relaxation thus developed is of prime importance for
the first experience of friendly otherness. One may say . . . that in
thus *getting what is given,* and in learning to *get somebody to do*
for him what he wishes to have done, the baby also develops the
necessary ego groundwork *to get to be* a giver."[6]

What these words of Erikson's make clear is that the initial
and crucial stages of human socialization implicate and enmesh the
totality of two organisms, the child and its mother, in an inseparable
mutuality of service to their deepest biological and emotional needs.
And out of this problematic reunification of mother and child—
problematic because it involves inevitably the trauma of separa-
tion—emerge the foundations of the human's knowledge of what it
means to give and to receive, to trust and to mistrust, to be a friend
and to have a friend, and to have a sense of hope and a sense of
doom.

Earlier, when speaking of theories (p. 140), I said that no
term of a theory can ever be fully and finally understood. We may
say the same thing about words generally, especially about such
words as trust and friendship and hope and their derivatives. Erikson teaches us that such words derive their meanings from universal, primal human experiences, and that any understanding of them
must always be fundamentally metaphoric. This profound truth also
informs us that man's entire understanding of his world, since it is
mediated by his language, must always and necessarily be bounded
by metaphoric descriptions. And since the child "meets the educa-
tive patterns of his culture," as Erikson says, "and thus learns the

basic modalities of human existence, each in personally and cultur-ally significant ways," each culture, indeed, each individual in a cul-ture, understands such words and language, hence the world, in a culturally and personally idiosyncratic way.

I could go on to describe the later stages of the socialization of the individual human, the effects of schooling, marriage, imprison-onment, warfare, hates and loves, the experiences of shame and guilt that vary so radically among the cultures of man, and so on. But that could be of no help to anyone who is not already convinced that any "understanding" a computer may be said to possess, hence any "in-telligence" that may be attributed to it, can have only the faintest relation to human understanding and human intelligence. We, how-ever, conclude that however much intelligence computers may at-tain, now or in the future, theirs must always be an intelligence *alien* to genuine human problems and concerns.

Still, the extreme or hardcore wing of the artificial intelligen-tsia will insist that the whole man, to again use Simon's expression, is after all an information processor, and that an information-proc-essing theory of man must therefore be adequate to account for his behavior in its entirety. We may agree with the major premise with-out necessarily drawing the indicated conclusion. We have already observed that a portion of the information the human "processes" is kinesthetic, that it is "stored" in his muscles and joints. It is simply not clear that such information, and the processing associated with it, can be represented in the form of computer programs and data structures at all.

It may, of course, be argued that it is in principle possible for a computer to simulate the entire network of cells that constitutes the human body. But that would introduce a theory of information processing entirely different from any which has so far been ad-vanced. Besides, such a simulation would result in "behavior" on such an incredibly long-time scale that no robot built on such prin-ciples could possibly interact with human beings. Finally, there ap-pears to be no prospect whatever that mankind will know enough neurophysiology within the next several hundred years to have the intellectual basis for designing such a machine. We may therefore dismiss such arguments.

There is, however, still another assumption that information-processing modelers of man make that may be false, and whose denial severely undermines their program: that there exists one and only one class of information processes, and that every member of that class is reducible to the kind of information processes exemplified by such systems as GPS and Schank-like language-understanding formalisms. Yet every human being has the impression that he thinks at least as much by intuition, hunch, and other such informal means as he does "systematically," that is by means such as logic. Questions like "Can a computer have original ideas? Can it compose a metaphor or a symphony or a poem?" keep cropping up. It is as if the folk wisdom knows the distinction between computer thought and the kind of thought people ordinarily engage in. The artificial intelligentsia, of course, do not believe there need be any distinction. They smile and answer "unproven."

Within the last decade or so, however, neurological evidence has begun to accumulate that suggests there may be a scientific basis to the folk wisdom.[7] It has long been known that the human brain consists of two so-called hemispheres that appear, superficially at least, to be identical. These two halves, which we will call LH (Left Hemisphere) and RH (Right Hemisphere), have, however, quite distinct functions. In righthanded people—and for simplicity, we can restrict our discussion to them—the LH may be said, at least roughly, to control the right half of the body, and the RH the left half. (Actually, the connectivities are somewhat more complex, particularly between the two brain halves and the eyes, but I will not go into such details here.) Most importantly, the two halves of the brain appear to have two quite distinct modalities of thought. The LH thinks, so to speak, in an orderly, sequential, and, we might call it, logical fashion. The RH, on the other hand, appears to think in terms of holistic images. Language processing appears to be almost exclusively centered in the LH, for example, whereas the RH is deeply involved in such tasks as spatial orientation, and the production and appreciation of music.

The distinct functions of the hemispheres of the brain began to be dramatically illustrated by patients who, after suffering from extremely severe forms of epilepsy, had their two brain halves surgically separated. In normal people, the two hemispheres are connected by a part of the brain called the corpus callosum. When this is cut, no direct communication between the two halves remains possible. It was found that when a so-called split-brain patient's hands were visually hidden from him and he was given, say, a pencil in his left hand, he could not *say* what had been given to him, but he could *show* that it was a pencil by drawing a picture of it or by selecting a picture of a pencil from among pictures of many different objects. However, when the experiment was repeated, only with the right hand receiving the pencil, then he could say it was a pencil but could not produce or recognize its pictorial representation. In the first situation, the RH received the "image" of the pencil and was able to encode it into pictorial representations, but not into linguistic structures. In the second, the LH received the "image" of the pencil and was able to encode it linguistically, but not pictorially.

There is also considerable evidence, which I will not detail here, that the RH is essentially the seat of intuition, and that it thinks quite independently of the LH. One way of characterizing intuitive thought is to say that, although it is logical, the standards of evidence it uses to make judgments are very different from the standards we normally associate with logical thought. In ordinary discourse, for example, when we say that two things are the same, we mean that they are identical in almost every respect; the standard of evidence we demand to justify such a judgment is extremely demanding. But when we construct a metaphor, e.g., the overseas Chinese are the Jews of the Orient, we pronounce two things to be the same in a very different sense. Metaphors are simply not logical; when taken literally, they are patently absurd. The RH, in other words, has criteria of absurdity that are far different from those of the logical LH.

The history of man's creativity is filled with stories of artists and scientists who, after working hard and long on some difficult problem, consciously decide to "forget" it, in effect, to turn it over to their RH. After some time, often with great suddenness and totally unexpectedly, the solution to their problem announces itself to them in almost complete form. The RH appears to have been able to overcome the most difficult logical and systematic problems by, I would conjecture, relaxing the rigid standards of thought of the LH. Given the looser standards the RH employs, it was perhaps able to design thought experiments which the LH simply could not, because of its rigidity, conceive. The RH is thus able to hit upon solutions which could then, of course, be recast into strictly logical terms by the LH. We may conjecture that in children the communication channel between the two brain halves is wide open; that is, that messages pass between the two halves quite freely. That may be why children are so incredibly imaginative; e.g., for them a cigar box is an automobile one moment and a house the next. In adults, the channel has been severely narrowed—whether by education or by physiological maturational processes or by both, I cannot guess. But it is clearly more open during the dream state. I may also conjecture that psychoanalysis, quite apart from its function as psychotherapy, trains people in the use of the channel. In psychoanalysis one learns, in Theodore Reik's happy phrase, to listen with the third ear, to attend, that is, to what the unconscious is "saying." Perhaps the various meditative disciplines serve the same purpose.

These are clearly conjectures, from which we are not entitled to draw any conclusions about how either humans or computers process information. Even as a mere possibility, however, they do raise a serious question about the universality of the mode of information processing we normally associate with logical thought and with computer programs.

That the right hemisphere of the brain is, loosely speaking, the "seat of intuition" is a hypothesis in favor of which evidence appears to be accumulating. Neither philosophers nor psychologists have yet been sufficiently persuaded by the existing evidence to confidently incorporate this hypothesis into their theories of mind. But this much is firmly established: the two hemispheres of the human brain think independently of one another; they think simultaneously; and they think in modes different from one another. Furthermore, we can say something about these two distinct modes.

The great mathematician Henri Poincaré, in his celebrated essay *Mathematical Creation*,[8] wrote

"The conscious self is narrowly limited, and as for the subliminal self we know not its limitations. . . . calculations . . . must be made in the . . . period of conscious work, that which follows the inspiration, that in which one verifies the results of this inspiration and deduces their consequences. The rules of these calculations are strict and complicated. They require discipline, attention, will, and therefore consciousness. In the subliminal self, on the contrary, reigns what I should call liberty, if we might give this name to the simple absence of discipline. . . . the privileged unconscious phenomena, those susceptible of becoming conscious, are those which, directly or indirectly, affect most profoundly our emotional sensibility. . . . The role of this unconscious work in mathematical invention appears to me incontestable, and traces of it would be found in other cases where it is less evident."

Of course, Poincaré, writing at the beginning of the twentieth century, knew nothing of the findings of the now-active brain researchers. And we are jumping to a conclusion when we identify what he calls the conscious and the subliminal selves with the left and right hemispheres of the brain, respectively. But our assertion here is that there are two distinct modes of human thought that operate independently and simultaneously. And that assertion Poincaré supports.

A most highly respected scientist who is now working, the psychologist Jerome Bruner, writes on this same topic from a slightly different perspective (recall that the right hand corresponds to the left hemisphere and the left hand to the right, or "intuitive," hemisphere):

"As a right-handed psychologist, I have been diligent for fifteen years in the study of the cognitive processes: how we acquire, retain, and transform knowledge of the world in which each of us lives—a world in part 'outside' us, in part 'inside.' The tools I have used have been those of the scientific psychologist studying perception, memory, learning, thinking, and (like a child of my times) I have addressed my inquiries to the laboratory rat as well as to human beings. At times, indeed, I have adopted the role of the clinician and carried out therapy with children. . . . There have been times when, somewhat discouraged by the complexities of the psychology of knowing, I have sought to escape through neurophysiology, to discover that the neurophysiologist can help only in the degree to which we can ask intelligent psychological questions of him.

"One thing has become increasingly clear in pursuing the nature of knowing. It is that the conventional apparatus of the psychologist—both his instruments of investigation and the conceptual tools he uses in the interpretation of his data—leaves one approach unexplored. It is an approach whose medium of exchange seems to be the metaphor paid out by the left hand. It is a way that grows happy hunches and 'lucky' guesses, that is stirred into connective activity by the poet and the necromancer looking sidewise rather than directly. Their hunches and intuitions generate a grammar of their own—searching out connections, suggesting similarities, weaving ideas loosely in a trial web. . . .

"[The psychologist] too searches widely and metaphorically for his hunches. He reads novels, looks at and even paints pictures, is struck by the power of myth, observes his fellow men intuitively and with wonder. In doing so, he acts only part-time like a proper psychologist, racking up cases against the criteria derived from hypothesis. Like his fellows, he observes the human scene with such sensibility as he can muster in the hope that his insight will be deepened. If he is lucky or if he has subtle psychological intuition, he will from time to time come up with hunches, combinatorial products of his metaphoric activity. If he is not fearful of these products of his own subjectivity, he will go so far as to tame the metaphors that have produced the hunches, tame them in the sense of shifting them from the left hand to the right hand by rendering them into notions that can be tested. It is my impression from observing myself and my colleagues that the forging of metaphoric hunch into testable hypothesis goes on all the time."[9]

That, of course, is my impression as well. Here Bruner speaks explicitly of the left hand, that is, the right hemisphere of the brain, as the artistic, the intuitive, and so on, and of the right hand, the left brain hemisphere, as the "conventional apparatus of the psychologist," and he speaks of the inadequacy of his "conceptual tools."

We learn from the testimony of hundreds of creative people as well as from our own introspection, that the human creative act always involves the conscious interpretation of messages coming from the unconscious, the shifting of ideas from the left hand to the right, in Bruner's phrase.

The unconscious is, of course, unconscious. It is like a seething, stormy sea within us. Its waves lap on the borders of our consciousness. And what we learn from it or about it, we construct from inferences we make about the meanings of the swells and surges, the breakers and ripples that wash the fringes of our consciousness. Occasionally we wander more deeply into the surf, as when we are in that semi-hypnagogic trance that divides sleep from wakefulness. But then we experience only chaos. Our thought modalities are maximally confused. And if we rip ourselves into waking, we cannot tell, we cannot translate or transform into linguistic modalities, what we had thought.

Does not the undoubted reality of this confusion, when placed alongside all the other available evidence to which I have alluded, lend weight to the altogether plausible conjecture that the forms of information manipulated in the right hemisphere of the brain, as well as the corresponding information processes, are simply different from those of the left hemisphere? And may it not be that we can in principle come to know those strange information forms and processes only in terms that are fundamentally irrelevant to the kind of understanding we seek? When, in the distant future, we come to know in detail how the brain functions on the neurophysiological level, we will, of course, be able to give an ultimately reductionist account of the functioning of the right hemisphere. But that would not be understanding in the sense we mean here, anymore than detailed knowledge of the electrical behavior of a running computer is, or even leads to, an understanding of the program the computer is running. On the other hand, a higher-level account of the functioning of the right hemisphere may always miss its most essential features, namely, those that differentiate it from the functioning of the left hemisphere. For we are constrained by our left-hemisphere thought modalities to always interpret messages coming from the right in left-hemisphere terms.

Perhaps the LH modality of thought is GPS-like, which is to say only that perhaps it can in principle be somehow formulated (not that GPS is even a candidate for a possible formulation). Perhaps it converts a problem like

> Tom has twice as many fish as Mary has guppies. If Mary has three guppies, how many fish does Tom have?

into its own terms, for example into

$$x = 2y; \, y = 3,$$

and solves it using information processing and symbol-manipulation techniques characteristic of GPS-like "thought." But it is then not possible for such a mechanism to have any idea of what fishes and guppies are, or of what it can mean to be a boy named Tom, and so on. Nor can the symbolic representation of the given problem be reconverted into the original problem statement. But human problem solving, perhaps even of the apparently most routine and mechanical variety, involves both left and right modes of thought. And certainly, direct human communication crucially involves the two hemispheres.

It is much too easy, especially for computer scientists, to be hypnotized by the "fact" that linguistic utterances are representable as linear strings of symbols. From this "fact" it is easy to deduce that linguistic communication is entirely a left-hemisphere affair. But human speech also has melody, and its song communicates as well as its libretto. Music is the province of the right hemisphere, as is the appreciation of gestures. As for written communication, its function is surely, at least in large part, to stimulate and excite especially the auditory imaginations of both the writer and the reader.

We may never know whether the conjecture that a part of us thinks in terms of symbolic structures that can be only sensed but not usefully explicated is true or false. Scientists, of course, abhor hypotheses that appear not to be falsifiable. Yet it may be that, under some profound conception of truth, the hypothesis is true. Perhaps it helps to explain why we remain lifelong strangers to ourselves and to each other, why every word in our lexicon is enveloped in at least some residual mystery, and why every attempt to solve life's problems by entirely rational means always fails.

But the inference that I here wish to draw from my conjecture is that, since we cannot know that it is false any more than that it is true, we are not entitled to the hubris so bombastically exhibited by the artificial intelligentsia. Even calculating reason compels the belief that we must stand in awe of the mysterious spectacle that is the whole man—I would even add, that is the whole ant.

There was a time when physics dreamed of explaining the whole of physical reality in terms of one comprehensive formalism. Leibnitz taught that if we knew the position and velocity of every elementary particle in the universe, we could predict the universe's whole future course. But then Werner Heisenberg proved that the very instruments man must use in order to measure physical phenomena disturb those phenomena, and that it is therefore impossible in principle to know both the exact position and the velocity of even a single elementary particle. He did not thereby falsify Leibnitz's conjecture. But he did show that its major premise was unattainable. That, of course, was sufficient to shatter the Leibnitzian dream. Only a little later, Kurt Gödel exposed the shakiness of the foundations of mathematics and logic itself by proving that every interesting formal system has some statements whose truth or falsity cannot be decided by the formal means of the system itself, in other words, that mathematics must necessarily be forever incomplete. It follows from this and others of Gödel's results that "The human mind is incapable of formulating (or mechanizing) all its mathematical intuitions. I.e.: If it has succeeded in formulating some of them, this very fact yields new intuitive knowledge."[10]

Both Heisenberg's so-called uncertainty principle and Gödel's incompleteness theorem sent terrible shock-waves through the worlds of physics, mathematics, and philosophy of science. But no one stopped working. Physicists, mathematicians, and philosophers more or less gracefully accepted the undeniable truth that there are limits to how far the world can be comprehended in Leibnitzian terms alone.

Much too much has already been made of the presumed implications of Heisenberg's and Gödel's results for artificial intelligence. I do not wish to contribute to that discussion here. But there is a sense in which psychology and artificial intelligence may usefully follow the example of the new-found humility of modern mathematics and physics: they should recognize that "while the constraints and limitations of logic do not exert their force on the things of the world, they do constrain and limit what are to count as defensible descriptions and interpretations of things."[11] Were they to recognize that, they could then take the next liberating step of also recognizing that truth is not equivalent to formal provability.

The lesson I have tried to teach here is not that the human mind is subject to Heisenberg uncertainties—though it may be—and that we can therefore never wholly comprehend it in terms of the kinds of reduction to discrete phenomena Leibnitz had in mind. The lesson here is rather that the part of the human mind which communicates to us in rational and scientific terms is itself an instrument that disturbs what it observes, particularly its voiceless partner, the unconscious, between which and our conscious selves it mediates. Its constraints and limitations circumscribe what are to constitute rational—again, if you will, scientific—descriptions and interpretations of the things of the world. These descriptions can therefore never be whole, anymore than a musical score can be a whole description or interpretation of even the simplest song.

But, and this is the saving grace of which an insolent and arrogant scientism attempts to rob us, we come to know and understand not only by way of the mechanisms of the conscious. We are capable of listening with the third ear, of sensing living truth that is truth beyond any standards of provability. It is *that* kind of understanding, and the kind of intelligence that is derived from it, which I claim is beyond the abilities of computers to simulate.

We have the habit, and it is sometimes useful to us, of speaking of man, mind, intelligence, and other such universal concepts. But gradually, even slyly, our own minds become infected with what A. N. Whitehead called the fallacy of misplaced concreteness. We come to believe that these theoretical terms are ultimately interpretable as observations, that in the "visible future" we will have ingenious instruments capable of measuring the "objects" to which these terms refer. There is, however, no such thing as mind; there are only individual minds, each belonging, not to "man," but to individual human beings. I have argued that intelligence cannot be measured by ingeniously constructed meter sticks placed along a one-dimensional continuum. Intelligence can be usefully discussed only in terms of domains of thought and action. From this I derive the conclusion that it cannot be useful, to say the least, to base serious work on notions of "how much" intelligence may be given to a computer. Debates based on such ideas—e.g., "Will computers ever exceed man in intelligence?"—are doomed to sterility.

I have argued that the individual human being, like any other organism, is defined by the problems he confronts. The human is unique by virtue of the fact that he must necessarily confront problems that arise from his unique biological and emotional needs. The human individual is in a constant state of becoming. The maintenance of that state, of his humanity, indeed, of his survival, depends crucially on his seeing himself, and on his being seen by other human beings, as a human being. No other organism, and certainly no computer, can be made to confront genuine human problems in human terms. And, since the domain of human intelligence is, except for a small set of formal problems, determined by man's humanity, every other intelligence, however great, must necessarily be alien to the human domain.

I have argued that there is an aspect to the human mind, the unconscious, that cannot be explained by the information-processing primitives, the elementary information processes, which we associate with formal thinking, calculation, and systematic rationality. Yet we are constrained to use them for scientific explanation, description, and interpretation. It behooves us, therefore, to remain aware of the poverty of our explanations and of their strictly limited scope. It is wrong to assert that any scientific account of the "whole man" is possible. There are some things beyond the power of science to fully comprehend.

The concept of an intelligence alien to certain domains of thought and action is crucial for understanding what are perhaps the most important limits on artificial intelligence. But that concept applies to the way humans relate to one another as well as to machines and their relation to man. For human socialization, though it is grounded in the biological constitution common to all humans, is strongly determined by culture. And human cultures differ radically among themselves. Countless studies confirm what must be obvious to all but the most parochial observers of the human scene: "The influence of culture is universal in that in some respects a man learns to become like all men; and it is particular in that a man who is reared in one society learns to become in some respects like all men of his society and not like those of others."[12] The authors of this quotation, students of Japanese society who lived among the Japanese for many years, go on to make the following observations:

> "In normal family life in Japan there is an emphasis on interdependence and reliance on others, while in America the emphasis is on independence and self-assertion. . . . In Japan the infant is seen more as a separate biological organism who from the beginning, in order to develop, needs to be drawn into increasingly interdependent relations with others. In America, the infant is seen more as a dependent biological organism who, in order to develop, needs to be made increasingly independent of others.
>
> "The Japanese baby seems passive, and he lies quietly with occasional unhappy vocalizations, while his mother, in her care, does more lulling, carrying, and rocking of her baby. She seems to try to soothe and quiet the child, and to communicate with him physically rather than verbally. On the other hand, the American infant is more active, happily vocal, and exploring of his environment, and his mother, in her care, does more looking at and chatting to her baby. She seems to stimulate the baby to activity and to vocal response. It is as if the American mother wanted to have a vocal, active baby, and the Japanese mother wanted to have a quiet, contented baby. In terms of styles of caretaking of the mothers in the two cultures, they get what they apparently want. . . . a great deal of cultural learning has taken place by three-to-four months of age. . . . babies have learned by this time to be Japanese and American babies in relation to the expectations of their mothers concerning their behavior.
>
> "[Adult] Japanese are more 'group' oriented and interdependent in their relations with others, while Americans are more 'individ-

ual' oriented and independent. . . . Japanese are more self-effacing and passive in contrast to Americans, who appear more self-assertive and aggressive. . . . Japanese are more sensitive to, and make conscious use of, many forms of nonverbal communication in human relations through the medium of gestures and physical proximity in comparison with Americans, who predominantly use verbal communication within a context of physical separateness.

"If these distinct patterns of behavior are well on the way to being learned by three-to-four months of age, and if they continue over the life span of the person, then there are very likely to be important areas of difference in emotional response in people of one culture when compared with those in another. Such differences are not easily subject to conscious control and, largely out of awareness, they accent and color human behavior. These differences . . . can also add to bewilderment and antagonism when people try to communicate across the emotional barriers of culture."[13]

Such profound differences in early training crucially affect the entire societies involved. And they are, of course, transmitted from one generation to the next and thus perpetuated. They must necessarily also help determine what members of the two societies know about their worlds, what are to be taken as "universal" cultural norms and values, hence what in each culture is and is not to be counted as fact. They determine, for example (and this is particularly relevant to the contrast between Japanese and American social norms), what are private as opposed to public conflicts, and hence what modes of adjudication are appropriate to the defense of what human interests. The Japanese traditionally prefer to settle disputes, even those for which relief at law is statutorily available, by what Westerners would see as informal means. Actually, these means are most often themselves circumscribed by stringent ritualistic requirements that are nowhere explicitly codified but are known to every Japanese of the appropriate social class. This sort of knowledge is acquired with the mother's milk and through the whole process of socialization that is itself so intimately tied to the individual's acquisition of his mother tongue. It cannot be learned from books; it cannot be explicated in any form but life itself.

An American judge, therefore, no matter what his intelligence and fairmindedness, could not sit in a Japanese family court. His intelligence is simply alien to the problems that arise in Japanese culture. The United States Supreme Court actively recognized this while it still had jurisdiction over distant territories. For example, in the case of Diaz v. Gonzales, which was originally tried in Puerto Rico, the court refused to set aside the judgment of the court of original jurisdiction, that is, of the native court. Justice Oliver W. Holmes, writing the opinion of the Court, stated,

"This Court has stated many times the deference due to understanding of the local courts upon matters of purely local concern. This is especially true when dealing with the decisions of a Court inheriting and brought up in a different system from that which prevails here. When we contemplate such a system from the outside it seems like a wall of stone, every part even with all the others, except so far as our own local education may lead us to see subordinations to which we are accustomed. But to one brought up within it, varying emphasis, tacit assumptions, unwritten practices, a thousand influences gained only from life, may give to the different parts wholly new values that logic and grammar never could have got from the books."[14]

Every human intelligence is thus alien to a great many domains of thought and action. There are vast areas of authentically human concern in every culture in which no member of another culture can possibly make responsible decisions. It is not that the outsider is unable to decide at all—he can always flip coins, for example—it is rather that the *basis* on which he would have to decide must be inappropriate to the context in which the decision is to be made.

What could be more obvious than the fact that, whatever intelligence a computer can muster, however it may be acquired, it must always and necessarily be absolutely alien to any and all authentic human concerns? The very asking of the question, "What does a judge (or a psychiatrist) know that we cannot tell a computer?" is a monstrous obscenity. That it has to be put into print at all, even for the purpose of exposing its morbidity, is a sign of the madness of our times.

Computers can make judicial decisions, computers can make psychiatric judgments. They can flip coins in much more sophisticated ways than can the most patient human being. The point is that they *ought* not be given such tasks. They may even be able to arrive at "correct" decisions in some cases—but always and necessarily on bases no human being should be willing to accept.

There have been many debates on "Computers and Mind." What I conclude here is that the relevant issues are neither technological nor even mathematical; they are ethical. They cannot be settled by asking questions beginning with "can." The limits of the applicability of computers are ultimately statable only in terms of oughts. What emerges as the most elementary insight is that, since we do not now have any ways of making computers wise, we ought not now to give computers tasks that demand wisdom.

9

INCOMPREHENSIBLE
PROGRAMS

We have in the preceding chapters seen something about
what computers are, where their power comes from, and how they
may be used for model building and for the embodiment of theories.
My concern has been with attempts to make computers behave intelligently, not with their application to mundane numerical problems. I have described and discussed some prominent research on
problem solving, on the simulation of cognitive processes, and on
natural-language understanding by computers, and have described
the visions of the artificial intelligentsia, often quoting its acknowledged leaders. I have hardly mentioned the failures that researchers
in artificial intelligence have suffered, because the failures that punc-
tuate every research effort are not necessarily grounds for despair; to
the contrary, they often enrich the soil from which better ideas later

spring. Besides, the fact that something has not yet been done, or even that an attempt to do it failed, does not demonstrate that it cannot be done. In attempting to avoid both trivial and sterile arguments—for example, arguments about whether computers can "in principle" be made to perform this or that specific task—I may even have created the impression that I think the potential (though not yet fully exploited) power of computers to be greater than, in fact, I believe it to be. In arguing that there are problems which confront man but which can never confront machines, and that man therefore comes to know things no machine can ever come to know, I may have created the impression that all problems that may confront both man and machine are potentially solvable by machines. Such is not my intention.

The achievements of the artificial intelligentsia are mainly triumphs of technique. They have contributed little either to cognitive psychology or to practical problem solving. To be sure, there have been what might be called spinoffs, such as refinements in higher-level programming languages, that were initiated by artificial-intelligence concerns and that have entered the mainstream of computer science. But these are hardly the results that the artificial intelligentsia has been forecasting for the "visible future" all these many years. With few exceptions, there have been no results, from over twenty years of artificial-intelligence research, that have found their way into industry generally or into the computer industry in particular.

Two exceptions are the remarkable programs **DENDRAL** and **MACSYMA** that exist at Stanford University and at M.I.T., respectively.[1] Both these programs perform highly technical functions whose discussion is far beyond the scope of this book. But a few words can be said about them.

DENDRAL interprets outputs of mass spectrometers, instruments used for analyses of chemical molecules. In ordinary practice, chemists in postdoctoral training are employed to deduce the chemical structures of molecules given to this instrument from the so-called mass spectra it produces. Their problem is somewhat analogous to that of reconstructing the life of a prehistoric village from the remains recovered by archeologists. There is, however, an important difference between the two problems: there exists a theory of mass spectrometry; that is, it is known how the instrument generates its output given a particular chemical for analysis. One can therefore evaluate a proffered solution by deducing from the theory what spectrum the instrument would produce if the chemical were what the tentative solution suggests it is. Unfortunately, limitations of precision intervene to make this process of inverse evaluation somewhat less than absolutely exact. Still, the analyst is in a better position than the archeologist, who has no strong methods for verifying his hypotheses. Stated in general terms, then, DENDRAL is a program that analyzes mass spectra and produces descriptions of the structures of molecules that, with very high probability, gave rise to these spectra. The program's competence equals or exceeds that of human chemists in analyzing certain classes of organic molecules.

MACSYMA is, by current standards, an enormously large program for doing symbolic mathematical manipulations. It can manipulate algebraic expressions involving formal variables, functions, and numbers. It can differentiate, integrate, take limits, solve equations, factor polynomials, expand functions in power series, and so on. It does all these things symbolically, not numerically. Thus, for example, given the problem of evaluating

$$\int \frac{dx}{a + bx},$$

it will produce

$$\frac{\log (a + bx)}{b}.$$

Of course, if it is given numerical values for all the variables involved, it will give the numerical value of the whole expression, but that is for it a relatively trivial task. Again, the technical details involved are beyond the scope of this discussion. What is important here is that, just as for DENDRAL, there exist strong theories about how the required transformations are to be made. Most importantly, especially for symbolic integration, it is possible (by differentiating) to check whether or not a proffered solution is in fact a solution, and, for integration, the test is absolute. And just as for DENDRAL, MACSYMA's task is one that is normally accomplished only by highly trained specialists.

These two programs owe a significant debt to the artificial-intelligence movement. They both use heuristic problem-solving methods in two distinct ways. First, when the design of these programs was initiated, the theories on which they are now based were not sufficiently well-formed to be modeled in terms of effective procedures. Yet people accomplished the required tasks. An initial problem was therefore to extricate from experts the heuristics they used in doing what they did. The initial versions of these programs were a mixture of algorithms incorporating those aspects of the problems that were well understood, and encodings of whatever heuristic techniques could be gleaned from experts. As the work progressed, however, the programs' heuristic components became increasingly well understood and hence convertible to enrichments of the relevant theories. Both programs were thus gradually modified until they became essentially completely theory-based. Second, heuristic methods were and continue to be used in both programs for reasons of efficiency. Both programs generate subproblems which, though in principle solvable by straightforward algorithmic means, yield more easily after they are classified as being amenable to solution by some special function and are then turned over to that function for solution. The development and refinement of both uses of heuristic methods, as well as many of the methods themselves, are products of artificial-intelligence research.

These two programs are distinguished from most other artificial-intelligence programs precisely in that they rest solidly on deep theories. The principal contributer of the theoretical underpinnings of DENDRAL was Joshua Lederberg, the geneticist and Nobel laureate, and MACSYMA's theoretical base is principally the work of Professor Joel Moses of M.I.T., an extremely talented and accomplished mathematician.

There are, of course, many other important and successful applications of computers. Computers, for example, control entire petroleum-refining plants, navigate spaceships, and monitor and largely control the environments in which astronauts perform their duties. Their programs rest on mathematical control theory and on firmly established physical theories. Such theory-based programs enjoy the enormously important advantage that, when they misbehave, their human monitors can detect that their performance does not correspond to the dictates of their theory and can diagnose the reason for the failure from the theory.

But most existing programs, and especially the largest and most important ones, are not theory-based in this way. They are heuristic, not necessarily in the sense that they employ heuristic methods internally, but in that their construction is based on rules of thumb, strategems that appear to "work" under most foreseen circumstances, and on other ad-hoc mechanisms that are added to them from time to time.

My own program, ELIZA, was of precisely this type. So is Winograd's language-understanding system and, all pretensions to the contrary notwithstanding, Newell and Simon's GPS. What is much more important, however, is that almost all the very large computer programs in daily use in industry, in government, and in the universities are of this type as well. These gigantic computer systems have usually been put together (one cannot always use the word "designed") by teams of programmers, whose work is often spread over many years. By the time these systems come into use, most of the original programmers have left or turned their attention to other pursuits. It is precisely when such systems begin to be used that their inner workings can no longer be understood by any single person or by a small team of individuals.

Norbert Wiener, the father of cybernetics, foretold this phenomenon in a remarkably prescient article published almost fifteen years ago. He said there;

> "It may well be that in principle we cannot make any machine
> the elements of whose behavior we cannot comprehend sooner or
> later. This does not mean in any way that we shall be able to

comprehend these elements in substantially less time than the time required for operation of the machine, or even within any given number of years or generations.

"An intelligent understanding of [a machine's] mode of performance may be delayed until long after the task which [it has] been set has been completed. . . . This means that, though machines are theoretically subject to human criticism, such criticism may be ineffective until long after it is relevant."[2]

What Norbert Wiener described as a possibility has long since become reality. The reasons for this appear to be almost impossible for the layman to understand or to accept. His misconception of what computers are, of what they do, and of how they do what they do is attributable in part to the pervasiveness of the mechanistic metaphor and the depth to which it has penetrated the unconscious of our entire culture. This is a legacy of the imaginative impact of the relatively simple machines that transformed life during the eighteenth and nineteenth centuries. It became "second nature" to virtually everyone living in the industrialized countries that to understand something was to understand it in mechanistic terms. Even great scientists of the late nineteenth century subscribed to this view. Lord Kelvin (1824–1907) wrote; "I never satisfy myself until I can make a mechanical model of a thing. If I can make a mechanical model, I can understand it. As long as I cannot make a mechanical model all the way through, I cannot understand it."[3] An expression of the corresponding modern sentiment is Minsky's belief that to understand music and "highly meaningful pictures" means to be able to write computer programs that can generate these things. But whereas Minsky deeply understands that computers are not machines to be equated with the mechanisms Kelvin knew, the layman understands just the contrary. To him computers and computer programs are "mechanical" in the same simple sense as steam engines and automobile transmissions.

This belief—and it is virtually universal among laymen—is reinforced by the slogan often repeated by computer scientists themselves that "unless a process is formulated with perfect precision, one cannot make a computer do it." This slogan is true, however, only under a very strict and most unusual interpretation of what it means to "formulate a process." If one were to throw a random pattern of bits into a computer's store, for example, and set the computer to interpret it as a program, then, assuming it would "work" at all, that bit pattern would be a "formulation" of some process. But program formulation is understood in normal discourse to mean that some agent (probably human) organized what is to become a computer program before giving it to the computer. The layman, having heard the slogan in question, believes that the very fact that a program runs on a computer guarantees that some programmer has formulated and understands every detail of the process which it embodies.

But his belief is contradicted by fact. A large program is, to use an analogy of which Minsky is also fond, an intricately connected network of courts of law, that is, of subroutines, to which evidence is transmitted by other subroutines. These courts weigh (evaluate) the data given them and then transmit their judgments to still other courts. The verdicts rendered by these courts may, indeed, often do, involve decisions about what court has "jurisdiction" over the intermediate results then being manipulated. The programmer thus cannot even know the path of decisionmaking within his own program, let alone what intermediate or final results it will produce. Program formulation is thus rather more like the creation of a bureaucracy than like the construction of a machine of the kind Lord Kelvin may have understood. As Minsky puts it:

"The programmer himself state[s] . . . 'legal' principles which permit . . . 'appeals,' he may have only a very incomplete understanding of when and where in the course of the program's operation these procedures will call on each other. And for a particular 'court,' he has only a sketchy idea of only some of the circumstances that will cause it to be called upon. In short, once past the beginner level, . . . programmers write—not 'sequences' [of instructions]—but specifications for the individuals of little societies. Try as he may he will often be unable fully to envision in advance all the details of their interactions. For that, after all, is why he needs the computer."[4]

Minsky goes on to make the following enormously important observations:

> "When a program grows in power by an evolution of partially understood patches and fixes, the programmer begins to lose track of internal details, loses his ability to predict what will happen, begins to hope instead of know, and watches the results as though the program were an individual whose range of behavior is uncertain.
>
> "This is already true in some big programs. . . . it will soon be much more acute. . . . large heuristic programs will be developed and modified by several programmers, each testing them on different examples from different [remotely located computer] consoles and inserting advice independently. The program will grow in effectiveness, but no one of the programmers will understand it all. (Of course, this won't always be successful—the interactions might make it get worse, and no one might be able to fix it again!) Now we see the real trouble with statements like 'it only does what its programmer told it to do.' There isn't any one programmer."[5]

We do not understand, to hark back to an earlier point for a moment, how a program of the kind Minsky here describes—one that, say, composes "great" music—helps us to "understand" music when the program itself is beyond our understanding.

But, more importantly, if the program has outrun the understanding of the agents who created it, what can it mean for it to "grow in effectiveness," or, for that matter, to "get worse"? As teachers (and we are all teachers) we, of course, constantly hope that those we instruct will grow in effectiveness in their dealings with whatever it is that is the subject of our tutelage. And we do not usually require of ourselves that we "understand" the processes whose growth we mean to encourage in our students, that is, that we understand them in the same way as, say, we understand the workings of a clock. Moreover, we do invest our hopes in our students and rely on them and trust them.

It is undoubtedly this kind of trust that Minsky urges us to invest in complex artificial-intelligence programs that grow in effectiveness but which come to be beyond our understanding. His advice is entirely reasonable if it applies to programs for which we have performance measures that enable us to tell, and tell in sufficient time, when these programs are operating outside an acceptable range of behavior or when, for any reason, they no longer deserve our trust. As we observed earlier, programs that are fundamentally models of well-understood theories fall into this class, even if, as may happen, no group of programmers commands a detailed understanding of the innards of the programs themselves. So too do programs whose drift away from performance criteria can be detected by observations of their moment-to-moment behavior—providing, of course, first, that someone responsible is watching, and, second, that he can intervene in time to avoid disaster. But there now exist many important programs that are very large and complex and that do not meet these criteria.

Minsky's account, which is entirely accurate, is thus of the utmost importance. It tells us that the condition Norbert Wiener described as a possibility in 1960 has quickly become and is now a reality. Minsky's words, moreover, take on a special importance because they were written by one of the chief architects and spokesmen for artificial intelligence, and were intended to correct the fuzzy thinking of humanists by extolling the power of computers, not their limitations.

Our society's growing reliance on computer systems that were initially intended to "help" people make analyses and decisions, but which have long since both surpassed the understanding of their users and become indispensable to them, is a very serious development. It has two important consequences. First, decisions are made with the aid of, and sometimes entirely by, computers whose programs no one any longer knows explicitly or understands. Hence no one can know the criteria or the rules on which such decisions are based. Second, the systems of rules and criteria that are embodied in such computer systems become immune to change, because, in the absence of a detailed understanding of the inner workings of a computer system, any substantial modification of it is very likely to render the whole system inoperative and possibly unrestorable. Such computer systems can therefore only grow. And their growth and the increasing reliance placed on them is then accompanied by an increasing legitimation of their "knowledge base."

Professor Philip Morrison of M.I.T. wrote a poignant parable on this theme:

"On the wall of my office is a world map, computer-plotted and therefore not as beautiful as draftsman would manage. On it are bold outlines, in eight or ten thousand dots, of the huge plates that make up the crust of the earth, which, when they spread apart or touch together or ride one over the other, generate most, perhaps nearly all, substantial earthquakes. The map embodies that realization, for its dotted outlines of plates were made by thousands of earthquake foci.

"The curious part is this: the seismologists responsible for the map say, somewhat apologetically, that since their own recordings of earthquakes were in one standard format which could easily be told to the machine so as to locate the dots on the map, they could use only their own data. They knew, to be sure, that seismology is much older than this decade, but the effort to try to connect the past to a standard coordinate system, to put in readable form into their computer the vast and diverse literature from 1840 until 1961—all this was beyond them. So they dropped out all reference to the science before 1961, and used only the earthquakes their own world-wide network of detectors recorded from 1961 to 1967. That, however, was as many as all the earthquakes recorded up to that time. They lost a factor of two, which is not much statistically; they gained the advantage of not having to read and interpret all those obscure German journals.

"This is a parable for the computer. Like all parables, it has an internal tension: it gives something to the enemies and to the friends of the computer alike. For the friends it is patent that this superb collection of epicenters delineating tectonic plates is probably the single greatest accomplishment of such synoptic study. For an outsider, it is fascinating to see the outline of the rifts and joints. At last we understand something of the earth in the large. At the same time, so cavalier a dismissal of the entire history of a science is breathtaking.

"The lesson is quite plain: nobody, not the most single-minded proponent of computer data processing, would say that it all began in 1961, even if our modern compatible data began then. The past was an indispensable prologue; it saw the formation of concepts, the development of techniques, the introduction of instruments, the idea of systematic recording, and so on. All this showed the way, without which I am sure the Coast and Geodetic Survey and its friends would not have been able to produce so beautiful a map."[6]

The computer has thus begun to be an instrument for the destruction of history. For when society legitimates only those "data" that are "in one standard format" and that "can easily be told to the machine," then history, memory itself, is annihilated. The *New York Times* has already begun to build a "data bank" of current events. Of course, only those data that are easily derivable as by-products of typesetting machines are admissable to the system. As the number of subscribers to this system grows, and as they learn more and more to rely on "all the news that [was once] fit to print," as the *Times* proudly identifies its editorial policy, how long will it be before what counts as fact is determined by the system, before all other knowledge, all memory, is simply declared illegitimate? Soon a supersystem will be built, based on the *New York Times'* data bank (or one very like it), from which "historians" will make inferences about what "really" happened, about who is connected to whom, and about the "real" logic of events. There are many people now who see nothing wrong in this.

We do not have to go to prospective systems to fill out Morrison's parable. In the recent American war against Viet Nam, computers operated by officers who had not the slightest idea of what went on inside their machines effectively chose which hamlets were to be bombed and what zones had a suffecent density of Viet Cong to be "legitimately" declared free-fire zones, that is, large geographical areas in which pilots had the "right" to kill every living thing. Of course, only "machine readable" data, that is, largely targeting information coming from other computers, could enter these machines. And when the American President decided to bomb Cambodia and to keep that decision secret from the American Congress, the computers in the Pentagon were "fixed" to transform the genuine strike reports coming in from the field into the false reports to which government leaders were given access. George Orwell's Ministry of Truth had become mechanized. History was not merely destroyed, it was recreated. And the high government leaders who felt themselves privileged to be allowed to read the secret reports that actually emerged from the Pentagon's computers of course believed them. After all, the computer itself had spoken. They did not realize that they had become their computer's "slaves," to use Admiral Moorer's own word, until the lies they instructed their computers to tell others ensnared them, the instructors, themselves.*

In modern warfare it is common for the soldier, say, the bomber pilot, to operate at an enormous psychological distance from his victims. He is not responsible for burned children because he never sees their village, his bombs, and certainly not the flaming children themselves. Modern technological rationalizations of war, diplomacy, politics, and commerce (such as computer games) have an even more insidious effect on the making of policy. Not only have policy makers abdicated their decision-making responsibility to a technology they do not understand—though all the while maintaining the illusion that they, the policy makers, are formulating policy questions and answering them—but responsibility has altogether evaporated. Not only does the most senior admiral of the United States Navy, in a rare moment of insight, perceive that he has become "a slave to these damned computers," that he cannot help but base his judgments on "what the computer says," but no human is responsible at all for the computer's output. The enormous computer systems in the Pentagon and their counterparts elsewhere in our culture have, in a very real sense, no authors. Thus they do not admit of any questions of right or wrong, of justice, or of any theory with which one can agree or disagree. They provide no basis on which "what the machine says" can be challenged. My father used

---

* According to a story in the *New York Times*, August 10, 1973, by Seymour Hersch, Admiral Thomas Moorer, chairman of the Joint Chiefs of Staff, explained to the U.S. Senate Armed Services Committee that air strikes against Cambodia were entered into the "Pentagon's large data computer" as strikes against South Viet Nam. The *Times*, as part of the same story, exhibits a photocopy of a strike report which carries the notation "All sorties targeted against Cambodia will be programmed against alternate targets in South Viet Nam." Admiral Moorer is reported to have said to the Senatorial Committee: "It is unfortunate that we had to become slaves to these damned computers."

to invoke the ultimate authority by saying to me "It stands written!" But then I read what stood written, imagine a human author, infer his values, and finally agree or disagree with him. Computer systems do not admit of exercises of imagination that may ultimately lead to authentic human judgment.

No wonder that men who live day in and day out with machines to which they perceive themselves to have become slaves begin to believe that men are machines, that, as an important scientist once put it:

> "It is possible to look on Man himself as a product of . . . an evolutionary process of developing robots, begotten by simpler robots, back to the primordial slime; . . . his ethical conduct [is] something to be interpreted in terms of the circuit action of . . . Man in his environment—a Turing machine with only two feedbacks determined, a desire to play and a desire to win."[7]

One would expect that large numbers of individuals, living in a society in which anonymous, hence irresponsible, forces formulate the large questions of the day and circumscribe the range of possible answers, would experience a kind of impotence and fall victim to a mindless rage. And surely we see that expectation fulfilled all around us, on university campuses and in factories, in homes and offices. Its manifestations are workers' sabotage of the products of their labor, unrest and aimlessness among students, street crime, escape into drug-induced dream worlds, and so on. Yet an alternative response is also very pervasive; as seen from one perspective, it appears to be resignation, but from another perspective it is what Erich Fromm long ago called "escape from freedom."

The "good German" in Hitler's time could sleep more soundly because he "didn't know" about Dachau. He didn't know, he told us later, because the highly organized Nazi system kept him from knowing. (Curiously, though, I, as an adolescent in that same Germany, knew about Dachau. I thought I had reason to fear it.) Of course, the real reason the good German didn't know is that he never felt it to be his responsibility to ask what had happened to his Jewish neighbor whose apartment suddenly became available. The university professor whose dream of being promoted to the status of

Ordinarius was suddenly fulfilled didn't ask how his precious chair had suddenly become vacant. Finally, all Germans became victims of what had befallen them.

Today even the most highly placed managers represent themselves as innocent victims of a technology for which they accept no responsibility and which they do not even pretend to understand. (One must wonder, though, why it never occurred to Admiral Moorer to ask what effect the millions of tons of bombs the computer said were being dropped on Viet Nam were having.) The American Secretary of State, Dr. Henry Kissinger, while explaining that he could hardly have known of the "White House horrors" revealed by the Watergate investigation, mourned over "the awfulness of events and the tragedy that has befallen so many people."

> "The tragedy so described had action, but no actors. Only 'events' were 'awful'—not individuals or officials. In this lifeless setting, the mockery of law and the deceit of the people had not been rehearsed and practiced: they had simply 'befallen.'"[8]

The myth of technological and political and social inevitability is a powerful tranquilizer of the conscience. Its service is to remove responsibility from the shoulders of everyone who truly believes in it.

But, in fact, there *are* actors!

For example, a planning paper circulated to the faculty and staff by the director of a major computer laboratory of a major university speaks as follows.[9]

> "Most of our research has been supported, and probably will continue to be supported, by the Government of the United States, the Department of Defense in particular. The Department of Defense, as well as other agencies of our government, is engaged in the development and operation of complex systems that have a very great destructive potential and that, increasingly, are commanded and controlled through digital computers. *These systems are responsible,* in large part, for the maintenance of what peace and stability there is in the world, and at the same time they are capable of unleashing destruction of a scale that is almost impossible for man to comprehend."

Note that systems are responsible, not people. Anyway, so much for a nod to their destructive potential; now on to the real concerns:

> "The crucial role of computers can be seen more vividly in military applications than in applications in the non-military sectors of society, but most of us have thought enough about the progressively increasing dependence upon computers in commerce and industry to project a picture in which the very functioning of society depends upon an orderly and meaningful execution of billions of electronic instructions every second. . . . there will be large-scale systems with millions of words of fast random-access memory, capable of tens of millions of instructions per second, in organizations of every kind. Most of these computers will be linked together in complexes of networks through which they will have access (governed by control mechanisms derived from what some of us are doing now) to all the information there is about everything and everybody. *And there is no stemming this trend in computer development.* . . .
>
> "In mastering the programming and control of computers, *we* especially could play a critical role. It may well be that no other organization is able to play this role as we are, yet no more important role may exist in science and engineering today.
>
> "The importance of the role stems, as has been noted, from the fact that *the computer has been incorporating itself,* and will surely continue to incorporate itself, into most of the functions that are fundamental to the support, protection, and development of our society. Even now, *there is no turning back,* and in a few years it will be clear that we are as vitally dependent upon the informational processing of our computers as upon the growth of grain in the field and the flow of fuel from the well."

There is not the slightest hint of a question as to whether we want this future. It is simply coming. We are helpless in the face of a tide that will, for no reason at all, not be stemmed. There is no turning back. Even the question is not worth discussing.

> "There are many facets to the general problem of mastery and control. They range from essentially philosophical problems that

concern meaning and intentions and the establishment of conformity between a plan and the actual behavior of a complex system to the almost purely technical problem of finding bugs in subroutines."

(In computer jargon, a "bug" is a programming error.) Notice the parochial, that is technological, view of philosophy that is displayed here. But to go on:

"One should not be satisfied with methods of programming that let bugs get into programs. It is probable that the best way to eliminate bugs is to devise bug-free methods of programming. Nevertheless, debugging [the elimination of bugs from programs] should be in the focus of the research effort undertaken to master programming. The reason is that research on debugging will yield insight into many problems in the formulation and expression of human intention. It is not the mere coding of a program formulation of problem and solution. As programming is mastered, there will be a continual enlarging of the scope of problem solving, a widening of the universe of discourse. The goal should be a method of programming that is as free of bugs and glitches in these higher levels as it is in the lower ones."

What is so remarkable about this is that the main—indeed, the only—impediment to "problem solving," even at "these higher levels," is seen to be entirely a matter of technical errors. There are no genuine conflicts in society. Once we understand "human intentions," itself a technical problem, all else is technique.

"Almost certainly, computers are freer of error in doing whatever they do than people are. If we can create a program-writing program, therefore, we should be a long step along the way to bug-free software. It is important, however, not to gloss over the problems that arise at the interface between the human statement of a problem and the computer understanding of that problem. The computer is unlikely to prepare a proper program unless its comprehension of the problem is entirely correct. . . .

"A possibly important approach to the mastering of programming and debugging is based on modeling. . . . In the beginning, the human programmer uses such models as he has available as aids. Toward the end, the models, combined into one comprehensive model, are doing most of the programming and debugging, but the human programmer is still in the picture to supervise or help out or provide heuristic guidance or whatever. Eventually, if the effort is successful, the model becomes the automatic programmer. . . .

"Only a year or two ago, it was necessary to put quotation marks around the word 'knowledge' whenever it was used in such a context as this . . . but [within a rather small circle of computer scientists] there is a consensus that we have reached the threshold beyond which one can think of computers as having knowledge and using it effectively and meaningfully in ways analogous, and probably in due course superior to, the ways in which human beings use knowledge. . . .

"Conversations with and among [our] faculty give rise to a strong feeling of convergence in a new direction. Many seem to sense a common set of priorities. Putting these feelings into words has required much sober thought, reflection and many extended discussions.

"The convergence of direction . . . involves making computers not only easy to use but, as has been stressed here, *trustworthy*. . . .

"[Our] unique resources are better utilized if totally harnessed to assure that the computer dominated future is one *we* wish. Perhaps [we] *only* [are] in a position to see this important goal attained."

The author of this document—he is, by the way, not anyone mentioned elsewhere in this book—is merely proposing the implementation of the program so often trumpeted by other spokesmen for technological optimism. What he writes is entirely consistent with, for example, H. A. Simon's forecast made in 1960 that

"Within the very near future—much less than twenty-five years—we shall have the technical capability of substituting machines for any and all human functions in organizations. Within the same period, we shall have acquired an extensive and empirically tested theory of human cognitive processes and their interaction with human emotions, attitudes, and values."*

Nor is the "optimism" displayed here related only to computers. Professor B. F. Skinner, the leader of behaviorism in psychology, and of whom it is often said that he is the most influential psychologist alive today, wrote recently,

> "The disastrous results of common sense in the management of human behavior are evident in every walk of life, from international affairs to the care of a baby, and we shall continue to be inept in all these fields until a scientific analysis clarifies the advantages of a more effective technology.
>
> "In the behavioristic view, man can now control his own destiny because he knows what must be done and how to do it."[10]

That last sentence cannot be read as meaning anything other than, "I, B. F. Skinner, know what must be done and how to do it," just as the last sentence of the planning paper I quoted cannot mean anything other than that the "we," whose job it is to harness our resources to assure the computer-dominated future that "we" wish, are the members of the rather small circle of computer scientists within which "we" can speak candidly and without the use of euphemistic quotation marks. It is significant that both sentences are the closing sentences of their documents. They contain, obviously, the final and most important message.

But the technological messiahs, who, because they find it impossible to trust the human mind, feel compelled to build "trustworthy" computers that will comprehend human intentions and

---

* H. A. Simon, "The Shape of Automation" (1960), reprinted in Z. W. Pylyshyn, ed., *Perspectives on the Computer Revolution* (Englewood Cliffs, N.J.: Prentice-Hall, 1970). In the same paper, Simon predicts: "Duplicating the problem-solving and information-handling capabilities of the brain is not far off; it would be surprising if it were not accomplished within the next decade." Well, more than a decade has passed, and the brain has remained as mysterious as ever. We must suppose that Prof. Simon is surprised. Sometimes, when watching, say, a particularly horrible scene in a movie, one manages not to be overcome by consciously reminding oneself that the people on the screen are, after all, only acting. That technique does not work here. Professor Simon is one of the most influential statesmen of science in America today. What he says really counts.

solve human problems, have competitors from other quarters as well. One of the most prominent among them is Professor J. W. Forrester of M.I.T., the intellectual father of the systems-dynamics movement. In testimony before a committee of the Congress of the United States,[11] he said,

> "It is my basic theme that the human mind is not adapted to interpreting how social systems behave. . . . Until recently there has been no way to estimate the behavior of social systems except by contemplation, discussion, argument, and guesswork."

In other words, the ways in which Plato, Spinoza, Hume, Mill, Gandhi, and so many others have thought about social systems are obviously inferior to the way of systems analysis. The trouble is that these ways of thinking are based on mental models. And

> "The mental model is fuzzy. It is incomplete. It is imprecisely stated. Furthermore, within one individual, a mental model changes with time and even during the flow of a single conversation. . . . Goals are different and are left unstated. It is little wonder that compromise takes so long."

Clearly, goals must be fixed, hence mental models too, else how can we determine the operators (to use GPS language) that are to be applied to the objects we wish to transform into "desired objects"? And the fuzziness of mental models is, Forrester observes, largely due to the fuzziness of human language itself. That must be repaired too.

> "Computer models differ from mental models in important ways. The computer models are stated explicitly. The 'mathematical' notation that is used for describing the model is unambiguous. It is a language that is clearer and more precise than the spoken languages like English or French. Computer model language is a simpler language. Its advantage is in the clarity of meaning and the simplicity of the language syntax. The language of a computer model can be understood by almost anyone, regardless of educational background. Furthermore any concept and relationship that can be clearly stated in ordinary language can be translated into computer model language."

One has to wonder why it is that ordinary language, what with all its dysfunctional properties, survives at all. And if it is so clear that every concept and relationship can be translated into computer terms, why do the linguists, e.g., Halle, Jakobson, Chomsky, continue to struggle so mightily? And why are there still poets? More to the present point, however, it is simply not true that "almost anyone" can understand the language of, say, Forrester's computer models. The latter have been widely accepted mainly because they were produced by a famous scientist affiliated with a prestigious university, and because their results are "what the computer says." Most ministers of state, labor leaders, and social commentators who have engaged in the "limits of growth" debate could no more read the computer programs which underly the controversy than they can read the equations of quantum physics. But, like Admiral Moorer, they find it useful to "trust" the machine.

Professor Forrester, finally, reassures his audience that "the means are visible" (to him, of course) for beginning to end uncertainty.

> "The great uncertainty with mental models is the inability to anticipate the consequences of interactions between parts of a system. This uncertainty is totally eliminated in computer models. Given a stated set of assumptions, the computer traces the resulting consequences without doubt or error."

He goes on to say that, although in our social system there are "no utopias" and no sustainable modes of behavior that are free of pressures and stresses, some possible modes of behavior are more "desirable" than others. And how are these more desirable modes of behavior enabled?

> "They seem to be possible only if we have a good understanding of the system dynamics and are willing to endure the self-discipline and pressures that must accompany the desirable mode."

There is undoubtedly some interpretation of the words "system" and "dynamics" which would lend a benign meaning to this observation. But in the context in which these words were spoken, they have the special meaning given them by Forrester. It is then clear that Forrester's message is quite the same as Skinner's and the others': the only way to gain the understanding which alone leads to "desirable modes of behavior" is Forrester-like (or Skinner-like, or GPS-like, and so on) methods of "scientific analysis."

The various systems and programs we have been discussing share some very significant characteristics: they are all, in a certain sense, simple; they all distort and abuse language; and they all, while disclaiming normative content, advocate an authoritarianism based on expertise. Their advocacy is, of course, disguised by their use of rhetoric couched in apparently neutral, jargon-laden, factual language (that is, by what the common man calls "bullshit"). These shared characteristics are, to some extent, separable, but they are not independent of one another.

The most superficial aspects of these systems' simplicity—as reflected by their simplistic construction of their subject matters—are immediately visible. Simon, for example, sees man as "quite simple." The "apparent" complexity of his behavior is due to the complexity of his environment. In any event, he can be simulated by a system sensitive to only "a few simple parameters," one that consists of only a few (certainly many fewer than, say, ten thousand) "elementary information processes." The laboratory director I quoted believes that the problem of human intentionality can be usefully attacked by research on computer program-debugging techniques, a belief shared by many of his colleagues. Skinner sees man as essentially a passive product (victim) of his genetic endowment and his history of reinforcing contingencies. The main difference between Skinner's system and GPS appears to be that Skinner is willing to look only at "input-output behavior" (to use computer jargon), whereas the architects of GPS and similar systems feel they can say something about what goes on inside the organism as well. But the philosophical differences between the two attitudes are slight. Forrester sees literally the whole world in terms of feedback loops.

> "Feedback loops are the fundamental building blocks of systems. . . . A feedback loop is composed of two kinds of variables,

called here rate and level variables. These two kinds of variables are necessary and sufficient. . . . the level variables are accumulations or integrations. . . . rates of flow cause the levels to change. The levels provide the information inputs to the rate equations which control the flows.

"The rate equations are the statements of system policy. They determine how the available information is converted to an action stream. . . . A rate equation states the discrepancy between the goal and the observed condition. And finally, the rate equation states the action that will result from the discrepancy."[12]

Notice the overlap in language between this statement and Newell and Simon's discussion of problems and problem solving (Chapter VI). The latter talk about "present objects" and "desired object," differences between them, operators that reduce these differences, goals, and so on. The difference between their system and Forrester's lies chiefly in the different sets of "elementary information processing" primitives each employs and, of course, in the fact that GPS uses heuristic methods to reduce searches for operators, etc., whereas in Forrester's system everything is explicitly algorithmrized in terms of rate and level variables enmeshed in feedback loops. But the worldviews represented by these two systems are basically the same. And they are very simple.

But these systems are simple in a deeper and more important sense as well. They have reduced reason itself to only its role in the domination of things, man, and, finally, nature.

"Concepts have been reduced to summaries of the characteristics that several specimens have in common. By denoting similarity, concepts eliminate the bother of enumerating qualities and thus serve better to organize the material of knowledge. They are thought of as mere abbreviations of the items to which they refer. Any use transcending auxiliary, technical summarization of factual data has been eliminated as a last trace of superstition. Concepts have become 'streamlined,' rationalized, labor-saving devices . . . thinking itself [has] been reduced to the level of industrial processes . . . in short, made part and parcel of production."[13]

No one who does not know the technical basis of the systems we have been discussing can possibly appreciate what a chillingly accurate account of them this passage is. It was written by the philosopher-sociologist Max Horkeimer in 1947, years before the forces that were even then eclipsing reason, to use Horkeimer's own expression, came to be embodied literally in machines.

This passage, especially in view of when and by whom it was written, informs us once again that the computer, as presently used by the technological elite, is not a cause of anything. It is rather an instrument pressed into the service of rationalizing, supporting, and sustaining the most conservative, indeed, reactionary, ideological components of the current *Zeitgeist*.

As we see so clearly in the various systems under scrutiny, meaning has become entirely transformed into function. Language, hence reason too, has been transformed into nothing more than an instrument for affecting the things and events in the world. Nothing these systems do has any intrinsic significance. There are only goals dictated by tides that cannot be turned back. There are only means-ends analyses for detecting discrepancies between the way things are, the "observed condition," and the way the fate that has befallen us tells us we wish them to be. In the process of adapting ourselves to these systems, we, even the admirals among us, have castrated not only ourselves (that is, resigned ourselves to impotence), but our very language as well. For now that language has become merely another tool, all concepts, ideas, images that artists and writers cannot paraphrase into computer-comprehensible language have lost their function and their potency. Forrester tells us this most clearly—but the others can be seen nodding their agreement: "Any concept and relationship that can be clearly stated in ordinary language can be translated into computer model language." The burden of proof that something has been "stated clearly" is on the poet. No wonder, given this view of language, that the distinction between the living and the lifeless, between man and machine, has become something less than real, at most a matter of nuance!

Corrupt language is very deeply imbedded in the rhetoric of the technological elite. We have already noted the transformation of the meaning of the word "understand" by Minsky into a purely instrumental term. And it is this interpretation of it that, of course, pervades all the systems we have been discussing. Newell and Simon's use of the word "problem" is another example and one just as significant.

During the times of trouble on American university campuses, one could often hear well-meaning speakers say that the unrest, at least on their campuses, was mainly caused by inadequate communication among the university's various constituencies, e.g., faculty, administration, students, staff. The "problem" was therefore seen as fundamentally a communication, hence a technical, problem. It was therefore solvable by technical means, such as the establishment of various "hotlines" to, say, the president's or the provost's office. Perhaps there were communication difficulties; there usually are on most campuses. But this view of the "problem"—a view entirely consistent with Newell and Simon's view of "human problem solving" and with instrumental reasoning—actively hides, buries, the existence of real conflicts. It may be, for example, that students have genuine ethical, moral, and political interests that conflict with interests the university administration perceives itself to have, and that each constituency understands the other's interests very well. Then there is a genuine problem, not a communication difficulty, certainly not one that can be repaired by the technical expedient of hotlines. But instrumental reason converts each dilemma, however genuine, into a mere paradox that can then be unraveled by the application of logic, by calculation. All conflicting interests are replaced by the interests of technique alone.

This, like Philip Morrison's story, is a parable too. Its wider significance is that the corruption of the word "problem" has brought in its train the mystique of "problem solving," with catastrophic effects on the whole world. When every problem on the international scene is seen by the "best and the brightest" problem solvers as being a mere technical problem, wars like the Viet Nam war become truly inevitable. The recognition of genuinely conflicting but legitimate interests of coexisting societies—and such recognition is surely a precondition to conflict resolution or accommodation—is rendered impossible from the outset. Instead, the simplest criteria are used to detect differences, to search for means to reduce these differences, and finally to apply operators to "present objects" in order to transform them into "desired objects." It is, in fact, entirely reasonable, if "reason" means instrumental reason, to apply American military force, B-52's, napalm, and all the rest, to "communist-dominated" Viet Nam (clearly an "undesirable object"), as the "operator" to transform it into a "desirable object," namely, a country serving American interests.

The mechanization of reason and of language has consequences far beyond any envisioned by the problem solvers we have cited. Horkeimer, long before computers became a fetish and gave concrete form to the eclipse of reason, gave us the needed perspective:

> "Justice, equality, happiness, tolerance, all the concepts that . . . were in preceding centuries supposed to be inherent in or sanctioned by reason, have lost their intellectual roots. They are still aims and ends, but there is no rational agency authorized to appraise and link them to an objective reality. Endorsed by venerable historical documents, they may still enjoy a certain prestige, and some are contained in the supreme law of the greatest countries. Nevertheless, they lack any confirmation by reason in its modern sense. Who can say that any one of these ideals is more closely related to truth than its opposite? According to the philosophy of the average modern intellectual, there is only one authority, namely, science, conceived as the classification of facts and the calculation of probabilities. The statement that justice and freedom are better in themselves than injustice and oppression is scientifically unverifiable and useless. It has come to sound as meaningless in itself as would the statement that red is more beautiful than blue, or that an egg is better than milk."[14]

As we ourselves have also observed, the reification of complex systems that have no authors, about which we know only that they were somehow given us by science and that they speak with its authority, permits no questions of truth or justice to be asked.

I cannot tell why the spokesmen I have cited want the developments they forecast to become true. Some of them have told me that they work on them for the morally bankrupt reason that "If we don't do it, someone else will." They fear that evil people will develop superintelligent machines and use them to oppress mankind, and that the only defense against these enemy machines will be superintelligent machines controlled by us, that is, by well-intentioned people. Others reveal that they have abdicated their autonomy by appealing to the "principle" of technological inevitability. But, finally, all I can say with assurance is that these people are not stupid. All the rest is mystery.

It is only a little easier to understand why the public embraces their ideas with at least equanimity and sometimes even with enthusiasm. The rhetoric of the technological intelligentsia may be attractive because it appears to be an invitation to reason. It is that, indeed. But, as I have argued, it urges instrumental reasonings, not authentic human rationality. It advertises easy and "scientifically" endorsed answers to all conceivable problems. It exploits the myth of expertise. Here too the corruption of language plays an important role. The language of the artificial intelligentsia, of the behavior modifiers, and of the systems engineers is mystifying. People, things, events are "programmed," one speaks of "inputs" and "outputs," of feedback loops, variables, parameters, processes, and so on, until eventually all contact with concrete situations is abstracted away. Then only graphs, data sets, printouts are left. And only "we," the experts, can understand them. "We" do—even if only to have good public relations—show our concern for the social consequences of "our" acts and plans. Planning papers, such as the one I quoted, almost always have an opening paragraph that makes a passing reference to the destructive potential of our instruments. And "we" do write essays on the social implications of our gadgetry. But, as I have remarked elsewhere, these pieces turn out to be remarkably self-serving.

"The structure of the typical essay on 'The impact of computers on society' is as follows: First there is an 'on the one hand' statement. It tells all the good things computers have already done for society and often even attempts to argue that the social order would already have collapsed were it not for the 'computer revolution.' This is usually followed by an 'on the other hand' caution, which tells of certain problems the introduction of computers

brings in its wake. The threat posed to individual privacy by large
data banks and the danger of large-scale unemployment induced
by industrial automation are usually mentioned. Finally, the glori-
ous present and prospective achievements of the computer are applauded, while the dangers alluded to in the second part are shown
to be capable of being alleviated by sophisticated technological
fixes. The closing paragraph consists of a plea for generous societal
support for more, and more large-scale, computer research and
development. This is usually coupled to the more or less subtle
assertion that only computer science, hence only the computer scientist, can guard the world against the admittedly hazardous fallout of applied computer technology."[15]

The real message of such typical essays is therefore that the expert
will take care of everything, even of the problems he himself creates.
He needs more money. That always. But he reassures a public that
does not want to know anyway.

    And what is the technologist's answer to such charges as are
here made?

    First of all, they are dismissed as being merely philosophical.
For example, my paper on the impact of computers on society drew
hundreds of letters, but only one from a member of the artificial-
intelligence community. It came from a former student of Prof. Simon, and said in part,

> "As far as society as a whole is concerned, the primary effects of
> computer technology are more important than their [sic] side ef-
> fects. It is only the more philosophically inclined who find the
> potential side effects more important. . . . It takes a rare person to
> spend more than a few hours pondering the philosophical implica-
> tions."[16]

This is, of course, entirely consistent with Horkeimer's observation
that language has lost even its right to speak in noninstrumental,
that is, philosophical, terms. But a more directly significant answer
was given by Dr. Kenneth B. Clark during a symposium held at
M.I.T. not long ago. He had just expressed his distress that M.I.T.
was not devoting more of its resources to the solution of social problems. He said (I quote from memory), "Here is a great institute

devoted to and expert in science and technology. Why, in this time of anguish, do you not apply your instruments, your techniques, to the burning social questions of the times?"

I suggested that answers to the urgent questions of the time might not be found exclusively in science and technology. I said that his own search for technological solutions to great problems, for example, his proposal to give tranquilizers to world leaders on a regular basis,[17] might be fundamentally misleading.

He responded, "I have very long ago come to the conclusion that answers to the great questions facing man at all times can come only from rational thinking. The only alternative is the kind of mindlessness that, as we have seen, leads only to violence and destruction."

One might very well endorse that view—but only if by rationality something other than the mere application of science and technology is meant, if rationality is not automatically and by implication equated to computability and to logicality. The alternative to the kind of rationality that sees the solution to world problems in psychotechnology is not mindlessness. It is reason restored to human dignity, to authenticity, to self-esteem, and to individual autonomy.

Instrumental reason has made out of words a fetish surrounded by black magic. And only the magicians have the rights of the initiated. Only they can say what words mean. And they play with words and they deceive us. When Skinner contrasts science with common sense and claims the first to be much superior, he means his "behavioral science" and he means the "common" in "common sense" pejoratively. He does not mean a common sense informed by a shared cultural perspective, or a common sense that, for no "reason" at all, balks at the idea that freedom and dignity are absurd and outmoded concepts.

The technologist argues again and again that views such as those expressed here are anti-technological, anti-scientific, and finally anti-intellectual. He will try to construe all arguments against his megalomanic visions as being arguments for the abandonment of reason, rationality, science, and technology, and in favor of pure intuition, feeling, drug-induced mindlessness, and so on. In fact, I *am* arguing for rationality. But I argue that rationality may not be separated from intuition and feeling. I argue for the rational use of science and technology, not for its mystification, let alone its abandonment. I urge the introduction of ethical thought into science planning. I combat the imperialism of instrumental reason, not reason.

It is said that men could always be found who thought that their own time was filled with the greatest forebodings of catastrophes to come, and even that theirs were the worst of all possible times for the whole of mankind. Certainly, we who were alive and awake during the time fascism appeared to be almost everywhere victorious saw the grim reaper making ready for civilization itself. Somehow civilization survived that threat—a threat that today's youth can no longer comprehend. But it cannot be said that civilization survived it, or the Great War that preceded it by only two decades, wholly intact. We came to know as never before what man can do to his fellows. Germany implemented the "final solution" of its "Jewish Problem" as a textbook exercise in instrumental reasoning. Humanity briefly shuddered when it could no longer avert its gaze from what had happened, when the photographs taken by the killers themselves began to circulate, and when the pitiful survivors re-emerged into the light. But in the end, it made no difference. The same logic, the same cold and ruthless application of calculating reason, slaughtered at least as many people during the next twenty years as had fallen victim to the technicians of the thousand-year Reich. We have learned nothing. Civilization is as imperiled today as it was then.

But if every time has heard the same Cassandra cry, then every time has also learned how little prophetic it seemed always to prove. Civilizations have been destroyed, many of them. But never mankind. But this time it is different. We are tired of hearing it, but we cannot deny it: this time man *is* able to destroy everything. Only his own decisions can save him.

It also used to be said that religion was the opiate of the people. I suppose that saying meant that the people were drugged with visions of the good life that would surely be theirs if they but patiently endured the earthly hell their masters made for them. On the other hand, it may be that religion was not addictive at all. Had it been, perhaps God would not have died and the new rationality would not have won out over grace. But instrumental reason, triumphant technique, and unbridled science *are* addictive. They create a concrete reality, a self-fulfilling nightmare. The optimistic technologists may yet be right: perhaps we have reached the point of no return. But why is the crew that has taken us this far cheering? Why do the passengers not look up from their games? Finally, now that we and no longer God are playing dice with the universe, how do we keep from coming up craps?

10

AGAINST THE IMPERIALISM
OF INSTRUMENTAL REASON

That man has aggregated to himself enormous power by means of his science and technology is so grossly banal a platitude that, paradoxically, although it is as widely believed as ever, it is less and less often repeated in serious conversation. The paradox arises because a platitude that ceases to be commonplace ceases to be perceived as a platitude. Some circles may even, after it has not been heard for a while, perceive it as its very opposite, that is, as a deep truth. There is a parable in that, too: the power man has acquired through his science and technology has itself been converted into impotence.

The common people surely feel this. Studs Terkel, in a monumental study of daily work in America, writes;

"For the many there is hardly concealed discontent. . . . 'I'm a machine,' says the spot welder. 'I'm caged,' says the bank teller, and echoes the hotel clerk. 'I'm a mule,' says the steel worker. 'A monkey can do what I do,' says the receptionist. 'I'm less than a farm implement,' says the migrant worker. 'I'm an object,' says the high fashion model. Blue collar and white call upon the identical phrase: 'I'm a robot.'"[1]

Perhaps the common people believe that, although they are powerless, there is power, namely, that exercised by their leaders. But we have seen that the American Secretary of State believes that events simply "befall" us, and that the American Chief of the Joint Chiefs of Staff confesses to having become a slave of computers. Our leaders cannot find the power either.

Even physicians, formerly a culture's very symbol of power, are powerless as they increasingly become mere conduits between their patients and the major drug manufacturers. Patients, in turn, are more and more merely passive objects on whom cures are wrought and to whom things are done. Their own inner healing resources, their capacities for self-reintegration, whether psychic or physical, are more and more regarded as irrelevant in a medicine that can hardly distinguish a human patient from a manufactured object. The now ascendant biofeedback movement may be the penultimate act in the drama separating man from nature; man no longer even senses himself, his body, directly, but only through pointer readings, flashing lights, and buzzing sounds produced by instruments attached to him as speedometers are attached to automobiles. The ultimate act of the drama is, of course, the final holocaust that wipes life out altogether.

Technological inevitability can thus be seen to be a mere element of a much larger syndrome. Science promised man power. But, as so often happens when people are seduced by promises of power, the price exacted in advance and all along the path, and the price actually paid, is servitude and impotence. Power is nothing if it is not the power to choose. Instrumental reason can make decisions, but there is all the difference between deciding and choosing.

The people Studs Terkel is talking about make decisions all day long, every day. But they appear not to make choices. They are, as they themselves testify, like Winograd's robot. One asks it "Why did you do that?" and it answers "Because this or that decision branch in my program happened to come out that way." And one asks "Why did you get to that branch?" and it again answers in the same way. But its final answer is "Because you told me to." Perhaps every human act involves a chain of calculations at what a systems engineer would call decision nodes. But the difference between a mechanical act and an authentically human one is that the latter terminates at a node whose decisive parameter is not "Because you told me to," but "Because I chose to." At that point calculations and explanations are displaced by truth. Here, too, is revealed the poverty of Simon's hypothesis that

> "The whole man, like the ant, viewed as a behaving system, is quite simple. The apparent complexity of his behavior over time is largely a reflection of the complexity of the environment in which he finds himself."

For that hypothesis to be true, it would also have to be true that man's capacity for choosing is as limited as is the ant's, that man has no more will or purpose, and, perhaps most importantly, no more a self-transcendent sense of obligation to himself as part of the continuum of nature, than does the ant. Again, it is a mystery why anyone would want to believe this to be the true condition of man.

But now and then a small light appears to penetrate the murky fog that obscures man's authentic capacities. Recently, for example, a group of eminent biologists urged their colleagues to discontinue certain experiments in which new types of biologically functional bacterial plasmids are created.[2] They express "serious concern that some of these artificial recombinant DNA molecules could prove biologically hazardous." Their concern is, so they write, "for the possible unfortunate consequences of the indiscriminate application of these techniques." Theirs is certainly a step in the right direction, and their initiative is to be applauded. Still, one may ask, why do they feel they have to give a reason for what they recommend at all? Is not the overriding obligation on men, including men of science, to exempt life itself from the madness of treating everything as an object, a sufficient reason, and one that does not even have to be spoken? Why does it have to be explained? It would appear that even the noblest acts of the most well-meaning people are poisoned by the corrosive climate of values of our time.

An easy explanation of this, and perhaps it contains truth, is that well-meaningness has supplanted nobility altogether. But there is a more subtle one. Our time prides itself on having finally achieved the freedom from censorship for which libertarians in all ages have struggled. Sexual matters can now be discussed more freely than ever before, women are beginning to find their rightful place in society, and, in general, ideas that could be only whispered until a decade or so ago may now circulate without restriction. The credit for these great achievements is claimed by the new spirit of rationalism, a rationalism that, it is argued, has finally been able to tear from man's eyes the shrouds imposed by mystical thought, religion, and such powerful illusions as freedom and dignity. Science has given to us this great victory over ignorance. But, on closer examination, this victory too can be seen as an Orwellian triumph of an even higher ignorance: what we have gained is a new conformism, which permits us to say anything that can be said in the functional languages of instrumental reason, but forbids us to allude to what Ionesco called the living truth. Just as our television screens may show us unbridled violence in "living color" but not scenes of authentic intimate love—the former by an itself-obscene reversal of values is said to be "real," whereas the latter is called obscene—so we may discuss the very manufacture of life and its "objective" manipulation, but we may not mention God, grace, or morality. Perhaps the biologists who urge their colleagues to do the right thing, but for the wrong reasons, are in fact motivated by their own deep reverence for life and by their own authentic humanity, only they dare not say so. In any case, such arguments would not be "effective," that is to say, instrumental.

If that is so, then those who censor their own speech do so, to use an outmoded expression, at the peril of their souls.

There is still another way to justify a scientist's renunciation of a particular line of research—and it is one from which all of us may derive lessons pertinent to our own lives. It begins from the principle that the range of one's responsibilities must be commensurate with the range of the effects of one's actions. In earlier times this principle led to a system of ethics that concerned itself chiefly with how persons conducted themselves toward one another. The biblical commandments, for example, speak mainly of what an individual's duties are toward his family and his neighbors. In biblical times few people could do anything that was likely to affect others beyond the boundaries of their own living spaces. Man's science and technology have altered this circumstance drastically. Not only can modern man's actions affect the whole planet that is his habitat, but they can determine the future of the entire human species. It follows therefore that man, particularly man the scientist and engineer, has responsibilities that transcend his immediate situation, that in fact extend directly to future generations. These responsibilities are especially grave since future generations cannot advocate their own cause now. We are all their trustees.[3]

The biologists' overt renunciation, however they themselves justify it, is an example which it behooves all scientists to emulate. Is this to suggest that scientists should close their minds to certain kinds of "immoral" hypotheses? Not at all. A scientific hypothesis is, at least from a scientific point of view, either true or false. This applies, for example, to Simon's hypotheses that man is "quite simple" and that he can be entirely simulated by a machine, as well as to McCarthy's hypothesis that there exists a logical calculus in terms of which all of reality can be formalized. It would be a silly error of logic to label such (or any other) hypotheses either moral or immoral or, for that matter, responsible or irresponsible.

But, although a scientific hypothesis can itself have no moral or ethical dimensions, an individual's decision to adopt it even tentatively, let alone to announce his faith in it to the general public, most certainly involves value judgments and does therefore have such dimensions. As the Harvard economist Marc J. Roberts recently wrote,

"Suppose we must choose between two hypotheses. No matter which we select, there is always the possibility that the other is correct. Obviously the relative likelihood of making a mistake when we select one or the other matters—but so too do the costs of alternative mistakes, the costs of assuming A is true when in fact B

is true or vice versa. We might well choose to risk a more likely small cost than a less likely large one. Yet the magnitude of the cost of being wrong in each case cannot be determined except on the basis of our values.

"Consider an extreme example: the view that there are genetic differences in the mental functioning of different races. Suppose society were to accept this view, and it proved false. I believe that very great evil would have been done. On the other hand, suppose society adopted the view that there are no differences, and that turned out to be incorrect. I would expect much less harm to result. Given these costs, I would want evidence which made the hypothesis of interracial similarity very unlikely indeed before I would reject it. My scientific choice depends on my values, not because I am uncritical or would like to believe that there are no such differences, but because consistent choices under uncertainty can only be made by looking at the cost of making alternative kinds of errors. In contrast, a would-be 'value-neutral scientist' would presumably be willing to operate on the assumption that such differences exist as soon as evidence made it even slightly more likely than the reverse assumption.

"These questions do not arise routinely in scientific work because traditional statistical methods typically subsume them under the choice of test criteria or of the particular technique to be used in estimating some magnitude. That choice is then made on conventional or traditional grounds, usually without discussion, justification, or even acknowledgement that value choices have been made."[4]

Roberts chose to illustrate that scientific hypotheses are not "value free" by citing the values enter into the scientist's choice to tolerate or not to tolerate the potential cost of being wrong. Values, as I will try to show, enter into choices made by scientists in other (and I believe even more important) ways as well. For the moment, however, I mean only to assert that it is entirely proper to say "bravo" to the biologists whose example we have cited, and to say "shame" to the scientists who recently wrote that "a machine-animal symbiont with an animal visual system and brain to augment mechanical functions" will be technically "feasible" within the next fifteen years.[5]

The introduction of words like "ethics" and "ought" into conversations about science seems almost always to engender a tension not unlike, I would say, the strain one can sense rising whenever, in conversation with elderly German university professors, one happens to allude to the career of one of their colleagues who prospered during the Hitler years. In the latter situation, the lowering of the social temperature betrays the fear that something "unfortunate" might be said, especially that the colleague's past inability to renounce his personal ambitions for the sake of morality might be mentioned. There is a recognition, then, of course, that the conduct not only of the colleague, but of all German academicians of the time, is in question. In the former situation, the tension betrays a similar concern, for ethics, at bottom, deals with nothing so much as renunciation. The tension betrays the fear that something will be said about what science, that is, scientists, ought and ought not to do. And there is a recognition that what might be talked about doesn't apply merely to science generally or to some abstract population known as scientists, but to the very people present.

Some scientists, though by no means all, maintain that the domain of science is universal, that there can be nothing which, as a consequence of some "higher" principle, ought not to be studied. And from this premise the conclusion is usually drawn that any talk of ethical "oughts" which apply to science is inherently subversive and anti-scientific, even anti-intellectual.

Whatever the merits of this argument as abstract logic may be, it is muddleheaded when applied to concrete situations, for there are infinitely many questions open to scientific investigation, but only finite resources at the command of science. Man must therefore choose which questions to attack and which to leave aside. We don't know, for example, whether the number of pores on an individual's skin is in any way correlated with the number of neurons in his brain. There is no interest in that question, and therefore no controversy about whether or not science ought to study it. The Chinese have practiced acupuncture for many centuries without arousing the interest of Western science. Now, suddenly, Western scientists have become interested. These examples illustrate that scientific "progress" does not move along some path determined by nature itself, but that it mirrors human interests and concerns.

Surely finely honed human intelligence is among the scarcest of resources available to modern society. And clearly some problems amenable to scientific investigation are more important than others. Human society is therefore inevitably faced with the task of wisely distributing the scarce resource that is its scientific talent. There simply is a responsibility—it cannot be wished away—to decide which problems are more important or interesting or whatever than others. Every specific society must constantly find ways to meet that responsibility. The question here is *how*, in an open society, these ways are to be found; are they to be dictated by, say, the military establishment, or are they to be open to debate among citizens and scientists? If they are to be debated, then why are ethics to be excluded from the discussion? And, finally, how can anything sensible emerge unless all first agree that, contrary to what John von Neuman asserted, technological possibilities are not irresistible to man? "Can" does not imply "ought."

Unfortunately, the new conformism that permits us to speak of everything except the few simple truths that are written in our hearts and in the holy books of each of man's many religions renders all arguments based on these truths—no matter how well thought out or eloquently constructed—laughable in the eyes of the scientists and technicians to whom they may be addressed. This in itself is probably the most tragic example of how an idea, badly used, turns into its own opposite. Scientists who continue to prattle on about "knowledge for its own sake" in order to exploit that slogan for their self-serving ends have detached science and knowledge from any contact with the real world. A central question of knowledge, once won, is its validation; but what we now see in almost all fields, especially in the branches of computer science we have been discussing, is that the validation of scientific knowledge has been reduced to the display of technological wonders. This can be interpreted in one of only two ways: either the nature to which science is attached consists entirely of raw material to be molded and manipulated as an object; or the knowledge that science has purchased for man is entirely irrelevant to man himself. Science cannot agree that the latter is true, for if it were, science would lose its license to practice. That loss would, of course, entail practical consequence (involving money and all that) which scientists would resist with all their might. If the former is true, then man himself has become an object. There is abundant evidence that this is, in fact, what has happened. But then knowledge too has lost the purity of which scientists boast so much; it has then become an enterprise no more or less important and no more inherently significant than, say, the knowledge of how to lay out an automobile assembly line. Who would want to know that "for its own sake"?

This development is tragic, in that it robs science of even the possibility of being guided by any authentically human standards, while it in no way restricts science's potential to deliver ever-increasing power to men. And here too we find the root of the much-talked-about dehumanization of man. An individual is dehumanized whenever he is treated as less than a whole person. The various forms of human and social engineering we have discussed here do just that, in that they circumvent all human contexts, especially those that give real meaning to human language.

The fact that arguments which appeal to higher principles—say, to an individual's obligations to his children, or to nature itself—are not acknowledged as legitimate poses a serious dilemma for anyone who wishes to persuade his colleagues to cooperate in imposing some limits on their research. If he makes such arguments anyway, perhaps hoping to induce a kind of conversion experience in his colleagues, then he risks being totally ineffective and even being excommunicated as a sort of comic fool. If he argues for restraint on the grounds that irreversible consequences may follow unrestrained research, then he participates in and helps to legitimate the abuse of instrumental reason (say, in the guise of cost-benefit analyses) against which he intends to struggle.

As is true of so many other dilemmas, the solution to this one lies in rejecting the rules of the game that give rise to it. For the present dilemma, the operative rule is that the salvation of the world—and that *is* what I am talking about—depends on converting others to sound ideas. That rule is false. The salvation of the world depends only on the individual whose world it is. At least, every individual must act as if the whole future of the world, of humanity itself, depends on him. Anything less is a shirking of responsibility and is itself a dehumanizing force, for anything less encourages the individual to look upon himself as a mere actor in a drama written by anonymous agents, as less than a whole person, and that is the beginning of passivity and aimlessness.

This is not an argument for solipsism, nor is it a counsel for every man to live only for himself. But it does argue that every man must live for himself first. For only by experiencing his own intrinsic worth, a worth utterly independent of his "use" as an instrument, can he come to know those self-transcendent ends that ultimately confer on him his identity and that are the only ultimate validators of human knowledge.

But the fact that each individual is responsible for the whole world, and that the discharge of that responsibility involves first of all each individual's responsibility to himself, does not deny that all of us have duties to one another. Chief among these is that we instruct one another as best we can. And the principal and most effective form of instruction we can practice is the example our own conduct provides to those who are touched by it. Teachers and writers have an especially heavy responsibility, precisely because they have taken positions from which their example reaches more than the few people in their immediate circle.

This spirit dictates that I must exhibit some of my own decisions about what I may and may not do in computer science. I do so with some misgivings, for I have learned that people are constantly asking one another what they must do, whereas the only really important question is what they must be. The physicist Steven Weinberg, in commenting on recent criticisms of science, writes, for example,

"I have tried to understand these critics by looking through some of their writings, and have found a good deal that is pertinent, and even moving. I especially share their distrust of those, from David Ricardo to the Club of Rome, who too confidently apply the methods of the natural sciences to human affairs. But in the end I am puzzled. What is it they want *me* to do?"[6]

My fear is that I will be understood to be answering a question of the kind Weinberg asks. That is not my intention. But the risk that I will be misunderstood cannot excuse me from my duty.

There is, in my view, no project in computer science as such that is morally repugnant and that I would advise students or colleagues to avoid. The projects I have been discussing, and others I will mention, are not properly part of computer science. Computers are not central to the work of Forrester and Skinner. The others are not computer science, because they are for the most part not science at all. They are, as I have already suggested, clever aggregations of techniques aimed at getting something done. Perhaps because of the accidents of history that caused academic departments whose concerns are with computers to be called "computer science" departments, all work done in such departments is indiscriminately called "science," even if only part of it deserves that honorable appellation. Tinkerers with techniques (gadget worshippers, Norbert Wiener called them) sometimes find it hard to resist the temptation to associate themselves with science and to siphon legitimacy from the reservoir it has accumulated. But not everyone who calls himself a singer has a voice.

Not all projects, by very far, that are frankly performance-oriented are dangerous or morally repugnant. Many really do help man to carry on his daily work more safely and more effectively. Computer-controlled navigation and collision-avoidance devices, for example, enable ships and planes to function under hitherto disabling conditions. The list of ways in which the computer has proved helpful is undoubtedly long. There are, however, two kinds of computer applications that either ought not be undertaken at all, or, if they are contemplated, should be approached with utmost caution.

The first kind I would call simply obscene. These are ones whose very contemplation ought to give rise to feelings of disgust in every civilized person. The proposal I have mentioned, that an animal's visual system and brain be coupled to computers, is an example. It represents an attack on life itself. One must wonder what must have happened to the proposers' perception of life, hence to their perceptions of themselves as part of the continuum of life, that they can even think of such a thing, let alone advocate it. On a much lesser level, one must wonder what conceivable need of man could be fulfilled by such a "device" at all, let alone by only such a device.

I would put all projects that propose to substitute a computer system for a human function that involves interpersonal respect, understanding, and love in the same category. I therefore reject Colby's proposal that computers be installed as psychotherapists, not on the grounds that such a project might be technically infeasible, but on the grounds that it is immoral. I have heard the defense that a person may get some psychological help from conversing with a computer even if the computer admittedly does not "understand" the person. One example given me was of a computer system designed to accept natural-language text via its typewriter console, and to respond to it with a randomized series of "yes" and "no." A troubled patient "conversed" with this system, and was allegedly led by it to think more deeply about his problems and to arrive at certain allegedly helpful conclusions. Until then he had just drifted in aimless worry. In principle, a set of Chinese fortune cookies or a deck of cards could have done the same job. The computer, however, contributed a certain aura—derived, of course, from science—that permitted the "patient" to believe in it where he might have dismissed fortune cookies and playing cards as instruments of superstition. The question then arises, and it answers itself, do we wish to encourage people to lead their lives on the basis of patent fraud, charlatanism, and unreality? And, more importantly, do we really believe that it helps people living in our already overly machine-like world to prefer the therapy administered by machines to that given by other people? I have heard this latter question answered with the assertion that my position is nothing more than "let them eat cake." It is said to ignore the shortage of good human psychotherapists, and to deny to troubled people what little help computers can now give them merely because presently available computers don't "yet"

measure up to, say, the best psychoanalysis. But that objection misses the point entirely. The point is (Simon and Colby to the contrary
notwithstanding) that there are some human functions for which
computers *ought* not to be substituted. It has nothing to do with
what computers can or cannot be made to do. Respect, understanding, and love are not technical problems.

The second kind of computer application that ought to be
avoided, or at least not undertaken without very careful forethought,
is that which can easily be seen to have irreversible and not entirely
foreseeable side effects. If, in addition, such an application cannot be
shown to meet a pressing human need that cannot readily be met in
any other way, then it ought not to be pursued. The latter stricture
follows directly from the argument I have already presented about
the scarcity of human intelligence.

The example I wish to cite here is that of the automatic
recognition of human speech. There are now three or four major
projects in the United States devoted to enabling computers to understand human speech, that is, to programming them in such a way
that verbal speech directed at them can be converted into the same
internal representations that would result if what had been said to
them had been typed into their consoles.

The problem, as can readily be seen, is very much more
complicated than that of natural-language understanding as such,
for in order to understand a stream of coherent speech, the language
in which that speech is rendered must be understood in the first
place. The solution of the "speech-understanding problem" therefore presupposes the solution of the "natural-language-understand-
ing problem." And we have seen that, for the latter, we have only
"the tiniest bit of relevant knowledge." But I am not here concerned
with the technical feasibility of the task, nor with any estimate of
just how little or greatly optimistic we might be about its comple-
tion.

Why should we want to undertake this task at all? I have
asked this question of many enthusiasts for the project. The most
cheerful answer I have been able to get is that it will help physicians
record their medical notes and then translate these notes into action
more efficiently. Of course, anything that has any ostensible connec-

tion to medicine is automatically considered good. But here we have to remember that the problem is so enormous that only the largest possible computers will ever be able to manage it. In other words, even if the desired system were successfully designed, it would probably require a computer so large and therefore so expensive that only the largest and best-endowed hospitals could possibly afford it—but in fact the whole system might be so prohibitively expensive that even they could not afford it. The question then becomes, is this really what medicine needs most at this time? Would not the talent, not to mention the money and the resources it represents, be better spent on projects that attack more urgent and more fundamental problems of health care?

But then, this alleged justification of speech-recognition "research" is merely a rationalization anyway. (I put the word "research" in quotation marks because the work I am here discussing is mere tinkering. I have no objection to serious scientists studying the psycho-physiology of human speech recognition.) If one asks such questions of the principal sponsor of this work, the Advanced Research Projects Agency (ARPA) of the United States Department of Defense, as was recently done at an open meeting, the answer given is that the Navy hopes to control its ships, and the other services their weapons, by voice commands. This project then represents, in the eyes of its chief sponsor, a long step toward a fully automated battlefield. I see no reason to advise my students to lend their talents to that aim.

I have urged my students and colleagues to ask still another question about this project: Granted that a speech-recognition machine is bound to be enormously expensive, and that only governments and possibly a very few very large corporations will therefore be able to afford it, what will they use it for? What can it possibly be used for? There is no question in my mind that there is no pressing human problem that will more easily be solved because such machines exist. But such listening machines, could they be made, will make monitoring of voice communication very much easier than it now is. Perhaps the only reason that there is very little government surveillance of telephone conversations in many countries of the world is that such surveillance takes so much manpower. Each conversation on a tapped phone must eventually be listened to by a human agent. But speech-recognizing machines could delete all "uninteresting" conversations and present transcripts of only the remaining ones to their masters. I do not for a moment believe that we will achieve this capability within the future so clearly visible to Newell and Simon. But I do ask, why should a talented computer technologist lend his support to such a project? As a citizen I ask, why should my government spend approximately 2.5 million dollars a year (as it now does) on this project?

Surely such questions presented themselves to thoughtful people in earlier stages of science and technology. But until recently society could always meet the unwanted and dangerous effects of its new inventions by, in a sense, reorganizing itself to undo or to minimize these effects. The density of cities could be reduced by geographically expanding the city. An individual could avoid the terrible effects of the industrial revolution in England by moving to America. And America could escape many of the consequences of the increasing power of military weapons by retreating behind its two oceanic moats. But those days are gone. The scientist and the technologist can no longer avoid the responsibility for what he does by appealing to the infinite powers of society to transform itself in response to new realities and to heal the wounds he inflicts on it. Certain limits have been reached. The transformations the new technologies may call for may be impossible to achieve, and the failure to achieve them may mean the annihilation of all life. No one has the right to impose such a choice on mankind.

I have spoken here of what ought and ought not to be done, of what is morally repugnant, and of what is dangerous. I am, of course, aware of the fact that these judgments of mine have themselves no moral force except on myself. Nor, as I have already said, do I have any intention of telling other people what tasks they should and should not undertake. I urge them only to consider the consequences of what they do do. And here I mean not only, not even primarily, the direct consequences of their actions on the world about them. I mean rather the consequences on themselves, as they construct their rationalizations, as they repress the truths that urge them to different courses, and as they chip away at their own autonomy. That so many people so often ask what they must do is a sign that the order of being and doing has become inverted. Those who know who and what they are do not need to ask what they should do. And those who must ask will not be able to stop asking until they begin to look inside themselves. But it is everyone's task to show by example what questions one can ask of oneself, and to show that one can live with what few answers there are.

But just as I have no license to dictate the actions of others, neither do the constructors of the world in which I must live have a right to unconditionally impose their visions on me. Scientists and technologists have, because of their power, an especially heavy responsibility, one that is not to be sluffed off behind a facade of slogans such as that of technological inevitability. In a world in which man increasingly meets only himself, and then only in the form of the products he has made, the makers and designers of these products—the buildings, airplanes, foodstuffs, bombs, and so on—need to have the most profound awareness that their products are, after all, the results of human choices. Men could instead choose to have truly safe automobiles, decent television, decent housing for everyone, or comfortable, safe, and widely distributed mass transportation. The fact that these things do not exist, in a country that has the resources to produce them, is a consequence, not of technological inevitability, not of the fact that there is no longer anyone who makes choices, but of the fact that people have chosen to make and to have just exactly the things we have made and do have.

It is hard, when one sees a particularly offensive television commercial, to imagine that adult human beings sometime and somewhere sat around a table and decided to construct exactly that commercial and to have it broadcast hundreds of times. But that is what happens. These things are not products of anonymous forces. They are the products of groups of men who have agreed among themselves that this pollution of the consciousness of the people serves their purposes.

But, as has been true since the beginning of recorded history, decisions having the most evil consequences are often made in the service of some overriding good. For example, in the summer of 1966 there was considerable agitation in the United States over America's intensive bombing of North Viet Nam. (The destruction rained on South Viet Nam by American bombers was less of an issue in the public debate, because the public was still persuaded that America was "helping" that unfortunate land.) Approximately forty American scientists who were high in the scientific estate decided to help stop the bombing by convening a summer study group under the auspices of the Institute of Defense Analyses, a prestigious consulting firm for the Department of Defense. They intended to demonstrate that the bombing was in fact ineffective.[7]

They made their demonstration using the best scientific tools, operations research and systems analysis and all that. But they felt they would not be heard by the Secretary of Defense unless they suggested an alternative to the bombing. They proposed that an "electronic fence" be placed in the so-called demilitarized zone separating South from North Viet Nam. This barrier was supposed to stop infiltrators from the North. It was to consist of, among other devices, small mines seeded into the earth, and specifically designed to blow off porters' feet but to be insensitive to truck passing over them. Other devices were to interdict truck traffic. The various electronic sensors, their monitors, and so on, eventually became part of the so-called McNamara line. This was the beginning of what has since developed into the concept of the electronic battlefield.

The intention of most of these men was not to invent or recommend a new technology that would make warfare more terrible and, by the way, less costly to highly industrialized nations at the expense of "underdeveloped" ones. Their intention was to stop the bombing. In this they were wholly on the side of the peace groups and of well-meaning citizens generally. And they actually accomplished their objective; the bombing of North Viet Nam was stopped for a time and the McNamara fence was installed. However, these enormously visible and influential people could have instead simply announced that they believed the bombing, indeed the whole American Viet Nam adventure, to be wrong, and that they would no longer "help." I know that at least some of the participants believed that the war was wrong; perhaps all of them did. But, as some of them explained to me later, they felt that if they made such an announcement, they would not be listened to, then or ever again.

Yet, who can tell what effect it would have had if forty of America's leading scientists had, in the summer of 1966, joined the peace groups in coming out flatly against the war on moral grounds? Apart from the positive effect such a move might have had on world events, what negative effect did their compromise have on themselves and on their colleagues and students for whom they served as examples?

There are several lessons to be learned from this episode. The first is that it was not technological inevitability that invented the electronic battlefield, nor was it a set of anonymous forces. Men just like the ones who design television commercials sat around a table and chose. Yet the outcome of the debates of the 1966 Summer Study were in a sense foreordained. The range of answers one gets is determined by the domain of questions one asks. As soon as it was settled that the Summer Study was to concern itself with only technical questions, the solution to the problem of stopping the bombing of the North became essentially a matter of calculation. When the side condition was added that the group must at all costs maintain its credibility with its sponsors, that it must not imperil the participants' "insider" status, then all degrees of freedom that its members might have had initially were effectively lost. Many of the participants have, I know, defended academic freedom, their own as well as that of colleagues whose careers were in jeopardy for political reasons. These men did not perceive themselves to be risking their scholarly or academic freedoms when they engaged in the kind of consulting characterized by the Summer Study. But the sacrifice of the degrees of freedom they might have had if they had not so thoroughly abandoned themselves to their sponsors, whether they made that sacrifice unwittingly or not, was a more potent form of censorship than any that could possibly have been imposed by officials of the state. This kind of intellectual self-mutilation, precisely because it is largely unconscious, is a principal source of the feeling of powerlessness experienced by so many people who appear, superficially at least, to occupy seats of power.

A second lesson is this. These men were able to give the counsel they gave because they were operating at an enormous psychological distance from the people who would be maimed and killed by the weapons systems that would result from the ideas they communicated to their sponsors. The lesson, therefore, is that the scientist and technologist must, by acts of will and of the imagination, actively strive to reduce such psychological distances, to counter the forces that tend to remove him from the consequences of his actions. He must—it is as simple as this—think of what he is actually doing. He must learn to listen to his own inner voice. He must learn to say "No!"

Finally, it is the act itself that matters. When instrumental reason is the sole guide to action, the acts it justifies are robbed of their inherent meanings and thus exist in an ethical vacuum. I recently heard an officer of a great university publicly defend an important policy decision he had made, one that many of the university's students and faculty opposed on moral grounds, with the words: "We could have taken a moral stand, but what good would that have done?" But the good of a moral act inheres in the act itself. That is why an act can itself ennoble or corrupt the person who performs it. The victory of instrumental reason in our time has brought about the virtual disappearance of this insight and thus perforce the delegitimation of the very idea of nobility.

I am aware, of course, that hardly anyone who reads these lines will feel himself addressed by them—so deep has the conviction that we are all governed by anonymous forces beyond our control penetrated into the shared consciousness of our time. And accompanying this conviction is a debasement of the idea of civil courage.

It is a widely held but a grievously mistaken belief that civil courage finds exercise only in the context of world-shaking events. To the contrary, its most arduous exercise is often in those small contexts in which the challenge is to overcome the fears induced by petty concerns over career, over our relationships to those who appear to have power over us, over whatever may disturb the tranquility of our mundane existence.

If this book is to be seen as advocating anything, then let it be a call to this simple kind of courage. And, because this book is, after all, about computers, let that call be heard mainly by teachers of computer science.

I want them to have heard me affirm that the computer is a powerful new metaphor for helping us to understand many aspects of the world, but that it enslaves the mind that has no other metaphors and few other resources to call on. The world is many things, and no single framework is large enough to contain them all, neither that of man's science nor that of his poetry, neither that of calculating reason nor that of pure intuition. And just as a love of music does not suffice to enable one to play the violin—one must also master the craft of the instrument and of music itself—so is it not enough to love humanity in order to help it survive. The teacher's calling to teach his craft is therefore an honorable one. But he must do more than that: he must teach more than one metaphor, and he must teach more by the example of his conduct than by what he writes on the blackboard. He must teach the limitations of his tools as well as their power.

It happens that programming is a relatively easy craft to learn. Almost anyone with a reasonably orderly mind can become a fairly good programmer with just a little instruction and practice. And because programming is almost immediately rewarding, that is, because a computer very quickly begins to behave somewhat in the way the programmer intends it to, programming is very seductive, especially for beginners. Moreover, it appeals most to precisely those who do not yet have sufficient maturity to tolerate long delays between an effort to achieve something and the appearance of concrete evidence of success. Immature students are therefore easily misled into believing that they have truly mastered a craft of immense power and of great importance when, in fact, they have learned only its rudiments and nothing substantive at all. A student's quick climb from a state of complete ignorance about computers to what appears to be a mastery of programming, but is in reality only a very minor plateau, may leave him with a euphoric sense of achievement and a conviction that he has discovered his true calling. The teacher, of course, also tends to feel rewarded by such students' obvious enthusiasm, and therefore to encourage it, perhaps unconsciously and against his better judgment. But for the student this may well be a trap. He may so thoroughly commit himself to what he naively perceives to be computer science, that is, to the mere polishing of his programming skills, that he may effectively preclude studying anything substantive.

Unfortunately, many universities have "computer science" programs at the undergraduate level that permit and even encourage students to take this course. When such students have completed their studies, they are rather like people who have somehow become eloquent in some foreign language, but who, when they attempt to write something in that language, find they have literally nothing of their own to say.

The lesson in this is that, although the learning of a craft is important, it cannot be everything.

The function of a university cannot be to simply offer prospective students a catalogue of "skills" from which to choose. For, were that its function, then the university would have to assume that the students who come to it have already become whatever it is they are to become. The university would then be quite correct in seeing the student as a sort of market basket, to be filled with goods from among the university's intellectual inventory. It would be correct, in other words, in seeing the student as an object very much like a computer whose storage banks are forever hungry for more "data." But surely that cannot be a proper characterization of what a university is or ought to be all about. Surely the university should look upon each of its citizens, students and faculty alike, first of all as human beings in search of—what else to call it?—truth, and hence in search of themselves. Something should constantly be happening to every citizen of the university; each should leave its halls having become someone other than he who entered in the morning. The mere teaching of craft cannot fulfill this high function of the university.

Just because so much of a computer-science curriculum is concerned with the craft of computation, it is perhaps easy for the teacher of computer science to fall into the habit of merely training. But, were he to do that, he would surely diminish himself and his profession. He would also detach himself from the rest of the intellectual and moral life of the university. The university should hold, before each of its citizens, and before the world at large as well, a vision of what it is possible for a man or a woman to become. It does this by giving ever-fresh life to the ideas of men and women who, by virtue of their own achievements, have contributed to the house we live in. And it does this, for better or for worse, by means of the example each of the university's citizens is for every other. The teacher of computer science, no more nor less than any other faculty member, is in effect constantly inviting his students to become what he himself is. If he views himself as a mere trainer, as a mere applier of "methods" for achieving ends determined by others, then he does his students two disservices. First, he invites them to become less than fully autonomous persons. He invites them to become mere followers of other people's orders, and finally no better than the machines that might someday replace them in that function. Second, he robs them of the glimpse of the ideas that alone purchase for computer science a place in the university's curriculum at all. And in doing that, he blinds them to the examples that computer scientists as creative human beings might have provided for them, hence of their very best chance to become truly good computer scientists themselves.[8]

Finally, the teacher of computer science is himself subject to the enormous temptation to be arrogant because his knowledge is somehow "harder" than that of his humanist colleagues. But the hardness of the knowledge available to him is of no advantage at all. His knowledge is merely less ambiguous and therefore, like his computer languages, less expressive of reality. The humanities particularly

> "have a greater familarity with an ambiguous, intractable, sometimes unreachable [moral] world that won't reduce itself to any correspondence with the symbols by means of which one might try to measure it. There is a world that stands apart from all efforts of historians to reduce [it] to the laws of history, a world which defies all efforts of artists to understand its basic laws of beauty. [Man's] practice should involve itself with softer than scientific knowledge. . . . that is not a retreat but an advance."[9]

The teacher of computer science must have the courage to resist the temptation to arrogance and to teach, again mainly by his own example, the validity and the legitimacy of softer knowledge. Why courage in this connection? For two reasons. The first and least important is that the more he succeeds in so teaching, the more he risks the censure of colleagues who, with less courage than his own, have succumbed to the simplistic worldviews inherent in granting imperial rights to science. The second is that, if he is to teach these things by his own example, he must have the courage to acknowledge, in Jerome Bruner's words, the products of his subjectivity.

Earlier I likened the unconscious to a turbulent sea, and the border dividing the conscious, logical mind from the unconscious to a stormy coastline. That analogy is useful here too. For the courage required to explore a dangerous coast is like the courage one must muster in order to probe one's unconscious, to take into one's heart and mind what it washes up on the shore of consciousness, and to examine it in spite of one's fears. For the unconscious washes up not only the material of creativity, not only pearls that need only be polished before being strung into structures of which one may then proudly speak, but also the darkest truths about one's self. These too must be examined, understood, and somehow incorporated into one's life.

If the teacher, if anyone, is to be an example of a whole person to others, he must first strive to be a whole person. Without the courage to confront one's inner as well as one's outer worlds, such wholeness is impossible to achieve. Instrumental reason alone cannot lead to it. And there precisely is a crucial difference between man and machine: Man, in order to become whole, must be forever an explorer of both his inner and his outer realities. His life is full of risks, but risks he has the courage to accept, because, like the explorer, he learns to trust his own capacities to endure, to overcome. What could it mean to speak of risk, courage, trust, endurance, and overcoming when one speaks of machines?

NOTES

Notes to Introduction

1. M. Polanyi, *The Tacit Dimension* (New York: Doubleday, Anchor ed., 1967), pp. 3-4.
2. This "conversation" is extracted from J. Weizenbaum, "**ELIZA**—A Computer Program For the Study of Natural Language Communication Between Man and Machine," *Communications of the Association for Computing Machinery*, vol. 9, no. 1 (January 1965), pp. 36-45.
3. K. M. Colby, J. B. Watt, and J. P. Gilbert, "A Computer Method of Psychotherapy: Preliminary Communication," *The Journal of Nervous and Mental Disease*, vol. 142, no. 2 (1966), pp. 148-152.
4. *Ibid.*
5. T. Winograd, "Procedures As A Representation For Data In A Computer Program For Understanding Natural Language." Ph.D. dissertation submitted to the Dept. of Mathematics (M.I.T.), August 24, 1970.
6. J. Weizenbaum, 1972.
7. Hubert L. Dreyfus, *What Computers Can't Do* (Harper and Row, 1972).
8. Hannah Arendt, *Crises of the Republic* (Harcourt Brace Jovanovich, Harvest edition, 1972), pp. 11 *et seq.*

Notes to Chapter 1

1. Alexander Marschak, *The Roots of Civilization* (New York: Macmillan, 1972), p. 57.
2. L. Mumford, *Technics and Civilization* (New York: Harcourt Brace Jovanovich, 1963), p. 14.
3. *Ibid*, pp. 13, 14.
4. *Ibid.*, p. 15.
5. Marschak, *op. cit.*, p. 14.
6. C. Pearson, *The Grammar of Science* (London: Dent, 1911), p. 11.
7. J. W. Forrester, in M. Greenberger, ed., *Managerial Decision Making in Management and the Computer of the Future* (Cambridge, Mass.: M.I.T. Press, 1962), pp. 52–53.

Notes to Chapter 2

1. Alan M. Turing, "On Computable Numbers, with an Application to the *Entscheidungsproblem,*" *Proc. London Math. Soc. Ser.* 2-42 (Nov. 17, 1936), pp. 230–265.
2. M. Minsky, *Computation, Finite and Infinite Machines* (Englewood Cliffs, N.J.: Prentice-Hall, 1967), p. 111.
3. M. Polanyi, *The Tacit Dimension* (New York: Doubleday, 1966), p. 4.

Note to Chapter 3

1. From J. Weizenbaum, "Contextual Understanding by Computers," *Communications of the ACM,* vol. 10, no. 8 (August 1967), pp. 474–480.

Notes to Chapter 4

1. Fyodor Dostoevski, *The Gambler,* quoted by E. Bergler, *The Psychology of Gambling* (New York: Hill and Wang, 1957), p. 33.
2. Bergler, *op. cit.*, p. 9.
3. *Ibid.*, p. 230.
4. *Ibid.*, p. 230.
5. M. Polanyi, *Personal Knowledge* (New York: Harper Torchbooks, 1964), p. 291.
6. *Ibid.*, p. 292.
7. Aldous Huxley, *Science, Liberty, and Peace* (New York: Harper, 1946), pp. 35–36.
8. H. A. Simon, *The Sciences of the Artificial* (Cambridge, Mass.: The M.I.T. Press, 1969), pp. 24–25.
9. Huxley, *op. cit.*, pp. 36–37.
10. Simon, *op. cit.*, pp. 52–53.

Notes to Chapter 5

1. For a brief and readily understandable discussion of Chomsky's position, see his *Problems of Knowledge and Freedom* (New York: Pantheon Books 1971), especially Chapter I. More complete and considerably more technical discussions are to be found in his *Aspects of the Structure of Syntax* (Cambridge, Mass.: The M.I.T. Press, 1965), and in the references there cited.

2. H. A. Simon and A. Newell, "Heuristic Problem Solving: The Next Advance in Operations Research," *Operations Research*, vol. 6 (Jan.–Feb. 1958), p. 8.

3. A. Kaplan, *The Conduct of Inquiry* (San Francisco, Calif.: Chandler, 1964), p. 296. This important book contains an excellent discussion of models and theories in the social sciences. See especially chapters VII and VIII.

4. K. B. Krauskopf and A. Besier, *Fundamentals in Physical Science* (New York: McGraw-Hill, 6th ed., 1971), p. 28.

5. Kaplan, *op. cit.*, p. 57.

6. A. Newell and H. A. Simon, *Human Problem Solving* (Englewood Cliffs, N.J.: Prentice-Hall, 1972), p. 10.

7. J. von Neuman, *The Computer and the Brain* (New Haven, Conn.: Yale University Press, 1958), p. 82.

8. P. Suppes, "Meaning and Uses of Models," in B. H. Kazemier and D. Vuysje, eds., *The Concept and the Role of the Model in Mathematics and Natural and Social Sciences* (New York: Gordon and Breach, 1961), p. 172.

Notes to Chapter 6

1. I. A. Richards, *The Philosophy of Rhetoric* (Oxford, England: Oxford University Press, 1936), p. 93.

2. Marvin Minsky, ed., *Semantic Information Processing* (Cambridge, Mass.: The M.I.T. Press, 1968), p. 12. From the introduction by M. Minsky.

3. G. A. Miller, *Language, Learning, and Models of the Mind.* Unpublished manuscript. June 1972.

4. *Ibid.*

5. S. Andreski, *Social Science as Sorcery* (New York: St. Martin's Press, 1972), p. 114.

6. Miller, *op. cit.*

7. Edward A. Feigenbaum, "The Simulation of Verbal Learning Behavior," in E. A. Feigenbaum and J. Feldman, eds., *Computers and Thought* (New York: McGraw-Hill, 1963), p. 299.

8. A. Newell, J. C. Shaw, and H. A. Simon, "Empirical Explorations of the Logic Theory Machine: A Case Study in Heuristics" (The RAND Corp., March 1957), Report P-951.

9. Hao Wang, "Toward Mechanical Mathematics," in K. M. Sayre and F. J. Coosson, eds., *The Modeling of Mind* (Notre Dame, Ind.: University of Notre Dame Press, 1963), pp. 91–120.

10. Newell *et al.*, *op. cit.*

11. G. Polya, *How to Solve It* (Copyright 1945 by Princeton University Press; © 1957 by G. Polya; Princeton Paperback 1971). A. Newell and H. A. Simon, *Human Problem Solving* (Englewood Cliffs, N.J.: Prentice-Hall, 1972).

12. Polya, *op. cit.*, pp. 129–133 (emphases are Polya's).

13. Newell and Simon, *op. cit.*, pp. 870–871.

14. H. A. Simon and A. Newell, "Heuristic Problem Solving: The Next Advance in Operations Research," *Operations Research*, vol. 6 (Jan.–Feb. 1958), pp. 8ff.

15. This system is described in some detail in Newell and Simon, *op. cit.*, particularly in Chapter 9, "Logic: GPS and Human Behavior," pp. 455–554.

16. *Ibid.*, p. 416.

17. *Ibid.*, pp. 864–866.

18. *Ibid.*, p. 809.

19. *Ibid.*, pp. 791, 809.

20. *Ibid.*, p. 5.

21. *Ibid.*, p. 855.

22. Simon and Newell, *op. cit.*, p. 8.

23. Newell and Simon, *Human Problem Solving*, p. 73.

24. H. A. Simon, "The Shape of Automation" (1960), reprinted in Z. W. Pylyshyn, ed., *Perspectives on the Computer Revolution* (Englewood Cliffs, N.J.: Prentice-Hall, 1970), p. 413.

25. K. M. Colby, "Simulations of Belief Systems," Chapter 6 in R. C. Schank and K. M. Colby, eds., *Computer Models of Thought and Language* (San Francisco, Calif.: W. H. Freeman and Co., 1973), p. 257.

26. K. M. Colby, J. B. Watt, and J. P. Gilbert, "A Computer Method of Psychotherapy: Preliminary Communication," *The Journal of Nervous and Mental Disease*, vol. 142, no. 2 (1966), pp. 148–152.

27. *Ibid.*, p. 150.

Notes to Chapter 7

1. R. K. Lindsay, "Inferential Memory as the Basis of Machines which Understand Natural Language," in E. A. Feigenbaum and J. Feldman, eds., *Computers and Thought* (New York: McGraw-Hill, 1963), p. 218.

2. B. F. Green, A. K. Wolf, C. Chomsky, and K. Laughery, "Baseball, An Automatic Question Answering System," in Feigenbaum and Feldman, *op. cit.*, pp. 207–216.

3. D. G. Bobrow, "Natural Language Input for a Computer Problem-Solving System," in M. Minsky, ed., *Semantic Information Processing* (Cambridge, Mass.: The M.I.T. Press, 1968), pp. 135–215.

4. J. Weizenbaum, "Contextual Understanding by Computers," *Communications of the ACM*, vol. 10, no. 8 (Aug. 1967), pp. 474–480.

5. See, however, Chapter III, p. 106. There ELIZA has a script that permits it to understand in a much stronger sense.

6. Roger C. Schank, "Identifications of Conceptualizations Underlying Natural Language," in R. C. Schank and K. M. Colby, eds., *Computer Models of Thought and Language* (San Francisco, Calif.: W. H. Freeman and Co., 1973), p. 191.

7. J. Weizenbaum, *op. cit.*, p. 476.

8. Terry Winograd, "A Procedural Model of Language Understanding," in Schank and Colby, *op. cit.*, p. 154.

9. *Ibid.*, p. 155f and p. 163.

10. *Ibid.*, p. 185.

11. G. A. Miller, *Language, Learning, and Models of the Mind* (unpublished manuscript, June 1972).

12. R. C. Schank, "Conceptual Dependency: A Theory of Natural Language Understanding," *Cognitive Psychology*, no. 3 (1972), pp. 553–554.
13. *Ibid.*, pp. 553, 629.

Notes to Chapter 8

1. In M. Greenberger, ed., *Management and the Computer of the Future* (Cambridge, Mass.: The M.I.T. Press, 1962), p. 123.
2. David C. McClelland, "Testing for Competence Rather Than for 'Intelligence,'"
American Psychologist, vol. 28, no. 1 (January 1973), pp. 1–14.
3. From Greenberger, *op. cit.*, p. 118.
4. See especially the work of R. A. Spitz, "Hospitalism," in *Psychoanalytic Study
of the Child*, vol. 1, 1945.
5. E. Erikson, *Childhood and Society* (New York: W. W. Norton, 2d ed., 1963), pp.
79, 80.
6. *Ibid.*, pp. 75–76.
7. For an account of the findings in this area, see Robert E. Ornstein, *The Psychology of Consciousness* (San Francisco, Calif.: W. H. Freeman and Co., 1972).
Chapter III is particularly relevant to the present discussion. It is written in
plain English. The references it cites open the door to the entire area of
research.
8. Reprinted in *The World of Mathematics* (New York: Simon and Schuster,
1956), vol. IV, pp. 2041–2050. This important essay is very much worth a trip
to the library, as is the set of volumes in which it appears.
9. J. Bruner, *On Knowing* (New York: Atheneum, 1973), pp. 3–5.
10. H. Wang, *From Mathematics to Philosophy* (New York: Humanities Press,
1974), p. 324. Kurt Gödel himself referred to this in December 1951 as one of
"the two most interesting rigorously proved results about minds and machines." The other is that either there exist certain formal questions which
neither humans nor machines can answer, or the human mind can answer
some formal questions that machines cannot.
11. D. C. Denett, "The Abilities of Men and Machines." Paper delivered to the
American Philosophical Association, December 29, 1970.
12. W. Caudill and H. Weinstein, "Maternal Care and Infant Behavior in Japan and
in America," *Psychiatry* 32 (1967): 12–43. Reprinted in C. S. Lavatelli and F.
Stendler, eds., *Readings in Child Behavior and Development* (New York: Harcourt
Brace Jovanovich, 3d ed., 1972), p. 78.
13. *Ibid.*, pp. 80 *et seq.*
14. Diaz v. Gonzales, 261 U.S. 102 (1923), Per Holmes, O. W. I owe this reference to
Professor Paul Freund of the Law School of Harvard University.

Notes to Chapter 9

1. On **DENDRAL**, see B. Buchanan, G. Sutherland, and E. A. Feigenbaum, "Heuristic **DENDRAL**: A Program for Generating Explanatory Hypotheses in Or-

ganic Chemistry," in B. Meltzer, ed., *Machine Intelligence* (New York: American Elsevier, 1969). On **MACSYMA,** see J. Moses, "Symbolic Integration, The Stormy Decade," in *Communications of the Assoc. for Computing Machinery,* vol. 14, no. 8 (1971), and W. A. Martin and R. J. Fateman, "The **MACSYMA** System" in *Proceedings of the Second Symposium on Symbolic and Algebraic Manipulation* (New York: Assoc. for Comp. Machinery, 1971).

2. N. Wiener, "Some Moral and Technical Consequences of Automation," *Science,* vol. 131, no. 3410 (May 6, 1960), p. 1355.

3. W. T. Kelvin, quoted by P. W. Bridgeman, in *The Logic of Modern Physics* (New York: Macmillan, 1946), p. 45.

4. M. Minsky, "Why Programming is a Good Medium for Expressing Poorly Understood and Sloppily Formulated Ideas," in M. Krampen and P. Seeitz, eds., *Design and Planning II* (New York: Hastings House, 1967), p. 120.

5. *Ibid.,* p. 121.

6. P. Morrison, "The Mind of the Machine," *Technology Review* (M.I.T.), January 1973, p. 13.

7. I will not supply the reference for this. It was written in 1952. When I knew the author many years later, he was a kind and humane old man. I cannot believe that he still subscribed to the view here cited in the last years before he died.

8. J. E. Hughes, "The Wide World of Watergate," in *Newsweek,* August 20, 1973, p. 13.

9. These extracts are quoted with the author's permission. I have added the emphases.

10. B. F. Skinner, *About Behaviorism* (New York: Alfred A. Knopf, 1974), pp. 234, 251.

11. The following quotations from Forrester are from "Testimony before the Subcommittee on Urban Growth of the Committee on Banking and Currency of the United States House of Representatives, given in Washington, D.C., October 7, 1970," 91st Congress, second session, part III, pp. 205–265. Available from the U.S. Government Printing Office.

12. J. W. Forrester, *Urban Dynamics* (Cambridge, Mass.: The M.I.T. Press, 1969), pp. 13ff.

13. M. Horkeimer, *Eclipse of Reason* (New York: Seabury, 1974), p. 21. This important book was first published by the Oxford University Press, N.Y., in 1947.

14. *Ibid.,* pp. 23–24.

15. J. Weizenbaum, "On the Impact of Computers on Society," *Science,* vol. 176, no. 12 (May 1972), pp. 609–614.

16. L. S. Coles, Letter to the editor, *Science,* November 10, 1972.

17. K. B. Clark, "Leadership and Psychotechnology," *N. Y. Times,* November 9, 1971, opposite the editorial page.

Notes to Chapter 10

1. S. Terkel, *Working* (New York: Pantheon, 1974), p. xi.

2. P. Berg *et al.,* letter to *Science,* July 26, 1974.

3. The point of this paragraph is argued at much greater length and very persuasively by Hans Jonas in his paper "Technology and Responsibility: Reflections on the New Tasks of Ethics," which appeared in *Social Research*, vol. 40, no. 1 (Spring 1973), pp. 31–54. I am grateful to Prof. Langdon Winner of M.I.T. for calling that paper to my attention.

4. Marc J. Roberts, "Nature and Condition of Social Science," in *Daedalus*, Summer 1974, pp. 54–55.

5. O. Firschein, M. A. Fischler, L. S. Coles, and J. M. Tenenbaum, "Intelligent Machines Are on the Way," *IEEE Spectrum*, July 1974, p. 43. The opinion I have quoted here appears to be the consensus of "41 artificial intelligence experts—including members of the International Joint Artificial Intelligence Council." The paper makes clear that the opinion is shared by the paper's authors.

6. Steven Weinberg, "Reflections of a Working Scientist," in *Daedalus*, Summer 1974, p. 41.

7. See the Gravel Edition of *The Pentagon Papers* (Boston, Mass.: Beacon Press, 1971), volume IV, especially p. 115.

8. Almost immediately after writing the last few paragraphs, I received a paper entitled "Methodology and the Study of Religion, Some Misgivings" from Wilfred Cantwell Smith, McCulloch Professor of Religion at Dalhousie University, Halifax, Nova Scotia. His paper expresses, among many other important ideas, many of the same points I have mentioned here. Since I had read some earlier papers by Prof. Smith, I cannot help but believe that I owe some of the ideas expressed here to him.

9. C. Oglesby, "A Juanist Way of Knowledge," lecture given to the M.I.T. Technology and Culture Seminar, Oct. 26, 1971. Copies are available from the Rev. John Crocker, Jr., 312 Memorial Drive, Cambridge, MA 02139.

INDEX